SpringerBriefs in Optimization

Series Editors

Panos M. Pardalos
János D. Pintèr
Stephen M. Robinson
Tamás Terlaky
My T. Thai

SpringerBriefs in Optimization showcases algorithmic and theoretical techniques, case studies, and applications within the broad-based field of optimization. Manuscripts related to the ever-growing applications of optimization in applied mathematics, engineering, medicine, economics, and other applied sciences are encouraged.

For further volumes:
http://www.springer.com/series/8918

Juan Peypouquet

Convex Optimization in Normed Spaces

Theory, Methods and Examples

 Springer

Juan Peypouquet
Universidad Técnica Federico
Valparaíso
Chile

ISSN 2190-8354   ISSN 2191-575X (electronic)
SpringerBriefs in Optimization
ISBN 978-3-319-13709-4   ISBN 978-3-319-13710-0 (eBook)
DOI 10.1007/978-3-319-13710-0

Library of Congress Control Number: 2015935427

Springer Cham Heidelberg New York Dordrecht London

Printed on acid-free paper

Springer is part of Springer Science+Business Media (www.springer.com)

To my family

Foreword

This book offers self-contained access and a modern vision of convex optimization in Banach spaces, with various examples and numerical methods. The challenge of providing a clear and complete presentation in only 120 pages is met successfully. This reflects a deep understanding by the author of this rich subject. Even when presenting classical results, he offers new insights and a fresh perspective.

Convex analysis lies in the core of optimization, since convexity is often present or hidden, even in apparently non-convex problems. It may be partially present in the basic blocks of structured problems, or introduced intentionally (as in relaxation) as a solution technique.

The examples considered in the book are carefully chosen. They illustrate a wide range of applications in mathematical programming, elliptic partial differential equations, optimal control, signal/image processing, among others. The framework of reflexive Banach spaces, which is analyzed in depth in the book, provides a general setting covering many of these situations.

The chapter on iterative methods is particularly accomplished. It gives a unified view on first-order algorithms, which are considered as time discrete versions of the (generalized) continuous steepest descent dynamic. Proximal algorithms are deeply studied. They play a fundamental role in splitting methods, which have been proven to be successful at reducing the dimension for large scale problems, like sparse optimization in signal/image processing, or domain decomposition for PDE's.

The text is accompanied by images that provide a geometric intuition of the analytical concepts. The writing is clear and concise, and the presentation is fairly linear. Overall, the book provides an enjoyable reading.

This book is accessible to a broad audience. It is excellent both for those who want to become familiar with the subject and learn the basic concepts, or as a textbook for a graduate course. The extensive bibliography provides the reader with more information, as well as recent trends.

Montpellier (France), October 2014   *Hedy Attouch*

Preface

Convex analysis comprises a number of powerful, elegant ideas, and techniques that are essential for understanding and solving problems in many areas of applied mathematics. In this introductory textbook, we have tried to present the main concepts and results in an easy-to-follow and mostly self-contained manner. It is intended to serve as a guide for students and more advanced independent learners who wish to get acquainted with the basic theoretical and practical tools for the minimization of convex functions on normed spaces. In particular, it may be useful for researchers working on related fields, who wish to apply convex-analytic techniques in their own work. It can also be used for teaching a graduate-level course on the subject, in view of the way the contents are organized. We should point out that we have not pretended to include *every* relevant result in the matter, but to present the ideas as transparently as possible. We believe that this work can be helpful to gain insight into some connections between optimization, control, calculus of variations, partial differential equations, and functional analysis that sometimes go unnoticed.

The book is organized as follows:

Keeping in mind users with different backgrounds, we begin by reviewing in Chap. 1 the minimal functional-analytic concepts and results convex analysis users should be familiar with. The chapter is divided into two sections: one devoted to normed spaces, including the basic aspects of differential calculus; and the other, containing the main aspects of Hilbert spaces.

Chapter 2 deals with existence of minimizers. We begin by providing a general result in a Hausdorff space setting, and then, we consider the case of convex functions defined on reflexive spaces. It is possible to go straight to the convex case but by working out the abstract, setting first one can deal with problems that are not convex but where convexity plays an important role.

Analysis and calculus are the core of Chap. 3. First, we discuss the connection between convexity, continuity, and differentiability. Next, we present the notion of subgradient, which extends the idea of derivative, along with its main properties and calculus rules. We use these concepts to derive necessary and sufficient optimality conditions for nonsmooth constrained optimization problems.

Chapter 4 contains selected applications: some functional analysis results are revisited under a convex-analytic perspective, existence of solutions as well as optimality conditions are established for optimal control and calculus of variations problems, and for some elliptic partial differential equations, including the obstacle problem and the p-Laplacian. We also mention the compressed sensing problem.

The main problem-solving strategies are described in Chap. 5. On the one hand, we provide a quick overview of the most popular discretization methods. On the other, we mention some ideas that help tackle the problems more easily. For instance, splitting is useful to reduce problem size or simplify optimality conditions, while Lagrangian duality and penalization allow us to pass from a constrained to an unconstrained setting.

In Chap. 6, standard abstract iterative methods are presented from a dynamical systems perspective and their main properties are discussed in detail. We finish by commenting how these methods can be applied in more realistically structured problems, by combining them with splitting, duality, and penalization techniques. We restrict ourselves to a Hilbert-space context in this chapter for two reasons: First, this setting is sufficiently rich and covers a broad range of situations. In particular, discretized problems, which are posed in finite-dimensional (Euclidean) spaces. Second, algorithms in Hilbert spaces enjoy good convergence properties that, in general, cannot be extended to arbitrary Banach spaces.

We intended to make this introductory textbook as complete as possible, but some relevant aspects of the theory and methods of convex optimization have not been covered. The interested reader is invited to consult related works. For a more general functional setting, see [10, 53, 105]. In [21], the authors did a remarkable job putting together essentially all there is to know about fixed point theory, convex analysis and monotone operators in Hilbert spaces, and elegantly pointing out the relationships. The exposition in [64] is deep, well-written, and rich in geometric insight (see also [26]). The timeless classic is, of course, [91]. For (convex and nonconvex) numerical optimization methods, see [24, 105]. Theoretical and practical aspects are clearly explained also in the two-volume textbook [66, 67]. More bibliographical commentaries are included throughout the text.

Finally, we invite the reader to check http://jpeypou.mat.utfsm.cl/ for additional material including examples, exercises, and complements. Suggestions and comments are welcome at juan.peypouquet@usm.cl.

Juan Peypouquet
Santiago & Valparaíso, Chile
October, 2014

Acknowledgments

This book contains my vision and experience on the theory and methods of convex optimization, which have been modeled by my research and teaching activity during the past few years, beginning with my PhD thesis at Universidad de Chile and Université Pierre et Marie Curie - Paris VI. I am thankful, in the first place, to my advisors, Felipe Álvarez and Sylvain Sorin, who introduced me to optimization and its connection to dynamical systems, and transmitted me their curiosity and passion for convex analysis. I am also grateful to my research collaborators, especially Hedy Attouch, who is always willing to share his deep and vast knowledge with kindness and generosity. He also introduced me to researchers and students at Université Montpellier II, with whom I have had a fruitful collaboration. Of great importance were the many stimulating discussions with colleagues at Universidad Técnica Federico Santa María, in particular, Pedro Gajardo and Luis Briceño, as well as my PhD students, Guillaume Garrigos and Cesare Molinari.

I also thank Associate Editor, Donna Chernyk, at Springer Science and Business Media, who invited me to engage in the more ambitious project of writing a book to be edited and distributed by Springer. I am truly grateful for this opportunity. A precursor [88] had been prepared earlier for a 20-h minicourse on convex optimization at the *XXVI EVM & EMALCA*, held in Mérida, in 2013. I thank my friend and former teacher at Universidad Simón Bolívar, Stefania Marcantognini, for proposing me as a lecturer at that School. Without their impulse, I would not have considered writing this book at this point in my career.

Finally, my research over the past few years was made possible thanks to Conicyt Anillo Project ACT-1106, ECOS-Conicyt Project C13E03, Millenium Nucleus ICM/FIC P10-024F, Fondecyt Grants 11090023 and 1140829, Basal Project CMM Universidad de Chile, STIC-AmSud Ovimine, and Universidad Técnica Federico Santa María, through their DGIP Grants 121123, 120916, 120821 and 121310.

Juan Peypouquet
Santiago & Valparaíso, Chile
October, 2014

Contents

Chapter 1
Basic Functional Analysis

Abstract Functional and convex analysis are closely intertwined. In this chapter
we recall the basic concepts and results from functional analysis and calculus that
will be needed throughout this book. A first section is devoted to general normed
spaces. We begin by establishing some of their main properties, with an emphasis
on the linear functions between spaces. This leads us to bounded linear functionals
and the topological dual. Second, we review the Hahn–Banach Separation Theorem, a very powerful tool with important consequences. It also illustrates the fact
that the boundaries between functional and convex analysis can be rather fuzzy at
times. Next, we discuss some relevant results concerning the weak topology, especially in terms of closedness and compactness. Finally, we include a subsection on
differential calculus, which also provides an introduction to standard smooth optimization techniques. The second section deals with Hilbert spaces, and their very
rich geometric structure, including the ideas of projection and orthogonality. We
also revisit some of the general concepts from the first section (duality, reflexivity,
weak convergence) in the light of this geometry.

For a comprehensive presentation, the reader is referred to [30] and [94].

1.1 Normed Spaces

A *norm* on a real vector space X is a function $\|\cdot\| : X \to \mathbf{R}$ such that

i) $\|x\| > 0$ for all $x \neq 0$;
ii) $\|\alpha x\| = |\alpha| \|x\|$ for all $x \in X$ and $\alpha \in \mathbf{R}$;
iii) The *triangle inequality* $\|x + y\| \leq \|x\| + \|y\|$ holds for all $x, y \in X$.

A *normed space* is a vector space where a norm has been specified.

Example 1.1. The space $\mathbf{R}^N$ with the norms: $\|x\|_\infty = \max\limits_{i=1,\dots,N} |x_i|$, or $\|x\|_p$
$= \left(\sum\limits_{i=1}^{N} |x_i|^p \right)^{1/p}$, for $p \geq 1$. □

© The Author(s) 2015
J. Peypouquet, *Convex Optimization in Normed Spaces*,
SpringerBriefs in Optimization, DOI 10.1007/978-3-319-13710-0_1

1

Example 1.2. The space $\mathscr{C}([a,b];\mathbf{R})$ of continuous real-valued functions on the interval $[a,b]$, with the norm $\|\cdot\|_\infty$ defined by $\|f\|_\infty = \max\limits_{t\in[a,b]} |f(t)|$. ☐

Example 1.3. The space $L^1(a,b;\mathbf{R})$ of Lebesgue-integrable real-valued functions on the interval $[a,b]$, with the norm $\|\cdot\|_1$ defined by $\|f\|_1 = \int_a^b |f(t)|\,dt$. ☐

Given $r > 0$ and $x \in X$, the *open ball* of radius r centered at x is the set

$$B_X(x,r) = \{y \in X : \|y-x\| < r\}.$$

The *closed ball* is

$$\bar{B}_X(x,r) = \{y \in X : \|y-x\| \le r\}.$$

We shall omit the reference to the space X whenever it is clear from the context. A set is *bounded* if it is contained in some ball.

In a normed space one can define a canonical topology as follows: a set V is a neighborhood of a point x if there is $r > 0$ such that $B(x,r) \subset V$. We call it the *strong topology*, in contrast with the *weak topology* to be defined later on.

We say that a sequence (x_n) in X *converges (strongly)* to $\bar{x} \in X$, and write $x_n \to \bar{x}$, as $n \to \infty$ if $\lim\limits_{n\to\infty} \|x_n - \bar{x}\| = 0$. The point $\bar{x}$ is the *limit* of the sequence. On the other hand, (x_n) has the *Cauchy property*, or it is a *Cauchy sequence*, if $\lim\limits_{n,m\to\infty} \|x_n - x_m\| = 0$. Every convergent sequence has the Cauchy property, and every Cauchy sequence is bounded. A *Banach space* is a normed space in which every Cauchy sequence is convergent, a property called *completeness*.

Example 1.4. The spaces in Examples 1.1, 1.2 and 1.3 are Banach spaces. ☐

We have the following result:

Theorem 1.5 (Baire's Category Theorem). *Let X be a Banach space and let (C_n) be a sequence of closed subsets of X. If each C_n has empty interior, then so does $\bigcup_{n\in\mathbf{N}} C_n$.*

Proof. A set $C \subset X$ has empty interior if, and only if, every open ball intersects C^c. Let B be an open ball. Take another open ball B' whose closure is contained in B. Since C_0^c has empty interior, $B' \cap C_0^c \neq \emptyset$. Moreover, since both B' and C_0^c are open, there exist $x_1 \in X$ and $r_1 \in (0,\tfrac{1}{2})$ such that $B(x_1,r_1) \subset B' \cap C_0^c$. Analogously, there exist $x_2 \in X$ and $r \in (0,\tfrac{1}{4})$ such that $B(x_2,r_2) \subset B(x_1,r_1) \cap C_1^c \subset B' \cap C_0^c \cap C_1^c$. Inductively, one defines $(x_m,r_m) \in X \times \mathbf{R}$ such that $x_m \in B(x_n,r_n) \cap \left(\bigcap_{k=0}^n C_k^c\right)$ and $r_m \in (0,2^{-m})$ for each $m > n \ge 1$. In particular, $\|x_m - x_n\| < 2^{-n}$ whenever $m > n \ge 1$. It follows that (x_n) is a Cauchy sequence and so, it must converge to some $\bar{x}$, which must belong to $\bar{B}' \cap \left(\bigcap_{k=0}^\infty C_k^c\right) \subset B \cap \left(\bigcup_{k=0}^\infty C_k\right)^c$, by construction. ☐

We shall find several important consequences of this result, especially the Banach–Steinhaus uniform boundedness principle (Theorem 1.9) and the continuity of convex functions in the interior of their domain (Proposition 3.6).

1.1.1 Bounded Linear Operators and Functionals, Topological Dual

Bounded Linear Operators

Let $(X, \|\cdot\|_X)$ and $(Y, \|\cdot\|_Y)$ be normed spaces. A linear *operator* $L : X \to Y$ is *bounded* if

$$\|L\|_{\mathscr{L}(X;Y)} := \sup_{\|x\|_X = 1} \|L(x)\|_Y < \infty.$$

The function $\|\cdot\|_{\mathscr{L}(X;Y)}$ is a norm on the space $\mathscr{L}(X;Y)$ of bounded linear operators from $(X, \|\cdot\|_X)$ to $(Y, \|\cdot\|_Y)$. For linear operators, boundedness and (uniform) continuity are closely related. This is shown in the following result:

Proposition 1.6. *Let* $(X, \|\cdot\|_X)$ *and* $(Y, \|\cdot\|_Y)$ *be normed spaces and let* $L : X \to Y$ *be a linear operator. The following are equivalent:*

i) L *is continuous in 0;*
ii) L *is bounded; and*
iii) L *is uniformly Lipschitz-continuous in X.*

Proof. Let i) hold. For each $\varepsilon > 0$ there is $\delta > 0$ such that $\|L(h)\|_Y \leq \varepsilon$ whenever $\|h\|_X \leq \delta$. If $\|x\|_X = 1$, then $\|L(x)\|_Y = \delta^{-1}\|L(\delta x)\|_Y \leq \delta^{-1}\varepsilon$ and so, $\sup_{\|x\|=1} \|L(x)\|_Y$
$< \infty$. Next, if ii) holds, then

$$\|L(x) - L(z)\|_Y = \|x - z\|_X \left\| L\left(\frac{x - z}{\|x - z\|_X} \right) \right\| \leq \|L\|_{\mathscr{L}(X;Y)} \|x - z\|_X$$

and L is uniformly Lipschitz-continuous. Clearly, iii) implies i). $\qquad\square$

We have the following:

Proposition 1.7. *If* $(X, \|\cdot\|_X)$ *is a normed space and* $(Y, \|\cdot\|_Y)$ *is a Banach space, then* $(\mathscr{L}(X;Y), \|\cdot\|_{\mathscr{L}(X;Y)})$ *is a Banach space.*

Proof. Let (L_n) be a Cauchy sequence in $\mathscr{L}(X;Y)$. Then, for each $x \in X$, the sequence $(L_n(x))$ has the Cauchy property as well. Since Y is complete, there exists $L(x) = \lim_{n \to \infty} L_n(x)$. Clearly, the function $L : X \to Y$ is linear. Moreover, since (L_n) is a Cauchy sequence, it is bounded. Therefore, there exists $C > 0$ such that $\|L_n(x)\|_Y \leq \|L_n\|_{\mathscr{L}(X;Y)} \|x\|_X \leq C\|x\|_X$ for all $x \in X$. Passing to the limit, we deduce that $L \in \mathscr{L}(X;Y)$ and $\|L\|_{\mathscr{L}(X;Y)} \leq C$. $\qquad\square$

The *kernel* of $L \in \mathscr{L}(X;Y)$ is the set

$$\ker(L) = \{x \in X : L(x) = 0\} = L^{-1}(0),$$

which is a closed subspace of X. The *range* of L is

$$R(L) = L(X) = \{L(x) : x \in X\}.$$

It is a subspace of Y, but it is not necessarily closed.

An operator $L \in \mathcal{L}(X;Y)$ is *invertible* if there exists an operator in $\mathcal{L}(Y;X)$, called the *inverse* of L, and denoted by L^{-1}, such that $L^{-1} \circ L(x) = x$ for all $x \in X$ and $L \circ L^{-1}(y) = y$ for all $y \in Y$. The set of invertible operators in $\mathcal{L}(X;Y)$ is denoted by $\mathrm{Inv}(X;Y)$. We have the following:

Proposition 1.8. *Let $(X, \|\cdot\|_X)$ and $(Y, \|\cdot\|_Y)$ be Banach spaces. The set $\mathrm{Inv}(X;Y)$ is open in $\mathcal{L}(X;Y)$ and the function $\Phi : \mathrm{Inv}(X;Y) \to \mathrm{Inv}(Y;X)$, defined by $\Phi(L) = L^{-1}$, is continuous.*

Proof. Let $L_0 \in \mathrm{Inv}(X;Y)$ and let $L \in B(L_0, \|L_0^{-1}\|^{-1})$. Let I_X be the identity operator in X and write $M = I_X - L_0^{-1} \circ L = L_0^{-1} \circ (L_0 - L)$. Denote by M^k the composition of M with itself k times and define $M_n = \sum_{k=0}^{n} M^k$. Since $\|M\| < 1$, (M_n) is a Cauchy sequence in $\mathcal{L}(X;X)$ and must converge to some $\bar{M}$. But $M \circ M_n = M_n \circ M = M_{n+1} - I_X$ implies $\bar{M} \circ (I_X - M) = (I_X - M) \circ \bar{M} = I_X$, which in turn gives

$$(\bar{M} \circ L_0^{-1}) \circ L = L \circ (\bar{M} \circ L_0^{-1}) = I_X.$$

It ensues that $L \in \mathrm{Inv}(X;Y)$ and $L^{-1} = \bar{M} \circ L_0^{-1}$. For the continuity, since

$$\|L^{-1} - L_0^{-1}\| \le \|L^{-1} \circ L_0 - I_X\| \cdot \|L_0^{-1}\| = \|L_0^{-1}\| \cdot \|\bar{M}^{-1} - I_X\|$$

and

$$\|M_{n+1} - I_X\| = \|M \circ M_n\| \le \|M\| \cdot (1 - \|M\|)^{-1},$$

we deduce that

$$\|L^{-1} - L_0^{-1}\| \le \frac{\|L_0^{-1}\|^2}{1 - \|M\|} \|L - L_0\|.$$

Observe that Φ is actually Lipschitz-continuous in every closed ball $\bar{B}(L_0, R)$ with $R < \|L_0^{-1}\|^{-1}$. $\square$

This fact will be used in Sect. 6.3.2. This and other useful calculus tools for normed spaces can be found in [33].

A remarkable consequence of linearity and completeness is that pointwise boundedness implies boundedness in the operator norm $\|\cdot\|_{\mathcal{L}(X;Y)}$. More precisely, we have the following consequence of Baire's Category Theorem 1.5:

Theorem 1.9 (Banach–Steinhaus uniform boundedness principle). *Let $(L_\lambda)_{\lambda \in \Lambda}$ be a family of bounded linear operators from a Banach space $(X, \|\cdot\|_X)$ to a normed space $(Y, \|\cdot\|_Y)$. If*

$$\sup_{\lambda \in \Lambda} \|L_\lambda(x)\|_Y < \infty$$

for each $x \in X$, then

$$\sup_{\lambda \in \Lambda} \|L_\lambda\|_{\mathcal{L}(X;Y)} < \infty.$$

Proof. For each $n \in \mathbf{N}$, the set

$$C_n := \{x \in X : \sup_{\lambda \in \Lambda} \|L_\lambda(x)\|_Y \leq n\}$$

is closed, as intersection of closed sets. Since $\cup_{n \in \mathbf{N}} C_n = X$ has nonempty interior, Baire's Category Theorem 1.5 shows the existence of $N \in \mathbf{N}$, $\hat{x} \in X$ and $\hat{r} > 0$ such that $B(\hat{x}, \hat{r}) \subset C_N$. This implies

$$r\|L_\lambda(h)\| \leq \|L_\lambda(\hat{x} + rh)\|_Y + \|L_\lambda(\hat{x})\|_Y \leq 2N$$

for each $r < \hat{r}$ and $\lambda \in \Lambda$. It follows that $\sup_{\lambda \in \Lambda} \|L_\lambda\|_{\mathscr{L}(X;Y)} < \infty$. $\square$

The Topological Dual and the Bidual

The *topological dual* of a normed space $(X, \|\cdot\|)$ is the normed space $(X^*, \|\cdot\|_*)$, where $X^* = \mathscr{L}(X; \mathbf{R})$ and $\|\cdot\|_* = \|\cdot\|_{\mathscr{L}(X;\mathbf{R})}$. It is actually a Banach space, by Proposition 1.7. Elements of $X*$ are *bounded linear functionals*. The bilinear function $\langle \cdot, \cdot \rangle_{X^*, X} : X^* \times X \to \mathbf{R}$, defined by

$$\langle L, x \rangle_{X^*, X} = L(x),$$

is the *duality product* between X and X^*. If the space can be easily guessed from the context, we shall write $\langle L, x \rangle$ instead of $\langle L, x \rangle_{X^*, X}$ to simplify the notation.

The *orthogonal space* or *annihilator* of a subspace V of X is

$$V^\perp = \{L \in X^* : \langle L, x \rangle = 0 \quad \text{for all} \quad x \in V\},$$

which is a closed subspace of X^*, even if V is not closed.

The *topological bidual* of $(X, \|\cdot\|)$ is the topological dual of $(X^*, \|\cdot\|_*)$, which we denote by $(X^{**}, \|\cdot\|_{**})$. Each $x \in X$ defines a linear function $\mu : X \to \mathbf{R}$ by

$$\mu_x(L) = \langle L, x \rangle_{X^*, X}$$

for each $L \in X^*$. Since $\langle L, x \rangle \leq \|L\|_* \|x\|$ for each $x \in X$ and $L \in X^*$, we have $\|\mu_x\|_{**} \leq \|x\|$, so actually $\mu_x \in X^{**}$. The function $\mathscr{J} : X \to X^{**}$, defined by $\mathscr{J}(x) = \mu_x$, is the *canonical embedding* of X into X^{**}. Clearly, the function $\mathscr{J}$ is linear and continuous. We shall see later (Proposition 1.17) that $\mathscr{J}$ is an isometry. This fact will imply, in particular, that the canonical embedding $\mathscr{J}$ is injective. The space $(X, \|\cdot\|)$ is *reflexive* if $\mathscr{J}$ is also surjective. In other words, if every element μ of X^{**} is of the form $\mu = \mu_x$ for some $x \in X$. Necessarily, $(X, \|\cdot\|)$ must be a Banach space since it is homeomorphic to $(X^{**}, \|\cdot\|_{**})$.

The Adjoint Operator

Let $(X, \|\cdot\|_X)$ and $(Y, \|\cdot\|_Y)$ be normed spaces and let $L \in \mathcal{L}(X;Y)$. Given $y^* \in Y^*$, the function $\zeta_{y^*} : X \to \mathbf{R}$ defined by

$$\zeta_{y^*}(x) = \langle y^*, Lx \rangle_{Y^*,Y}$$

is linear and continuous. In other words, $\zeta_{y^*} \in X^*$. The *adjoint* of L is the operator $L^* : Y^* \to X^*$ defined by

$$L^*(y^*) = \zeta_{y^*}.$$

In other words, L and L^* are linked by the identity

$$\langle L^* y^*, x \rangle_{X^*,X} = \langle y^*, Lx \rangle_{Y^*,Y}$$

We immediately see that $L^* \in \mathcal{L}(Y^*;X^*)$ and

$$\|L^*\|_{\mathcal{L}(Y^*;X^*)} = \sup_{\|y^*\|_{Y^*}=1} \left[\sup_{\|x\|_X=1} \langle y^*, Lx \rangle_{Y^*,Y} \right] \leq \|L\|_{\mathcal{L}(X;Y)}.$$

We shall verify later (Corollary 1.18) that the two norms actually coincide.

1.1.2 The Hahn–Banach Separation Theorem

The Hahn–Banach Separation Theorem is a cornerstone in functional and convex analysis. As we shall see in next chapter, it has several important consequences.

Let X be a real vector space. A set $C \subseteq X$ is *convex* if for each pair of points of C, the segment joining them also belongs to C. In other words, if the point $\lambda x + (1-\lambda)y$ belongs to C whenever $x, y \in C$ and $\lambda \in (0,1)$.

Theorem 1.10 (Hahn–Banach Separation Theorem). *Let A and B be nonempty, disjoint convex subsets of a normed space $(X, \|\cdot\|)$.*

i) *If A is open, there exists $L \in X^* \setminus \{0\}$ such that $\langle L, x \rangle < \langle L, y \rangle$ for each $x \in A$ and $y \in B$.*
ii) *If A is compact and B is closed, there exist $L \in X^* \setminus \{0\}$ and $\varepsilon > 0$ such that $\langle L, x \rangle + \varepsilon \leq \langle L, y \rangle$ for each $x \in A$ and $y \in B$.*

Remarks

Before proceeding with the proof of the Hahn–Banach Sepatarion Theorem 1.10, some remarks are in order:

First, Theorem 1.10 is equivalent to

Theorem 1.11. *Let C be a nonempty, convex subset of a normed space* $(X, \|\cdot\|)$ *not containing the origin.*

i) *If C is open, there exists* $L \in X^* \setminus \{0\}$ *such that* $\langle L, x \rangle < 0$ *for each* $x \in C$.
ii) *If C is closed, there exist* $L \in X^* \setminus \{0\}$ *and* $\varepsilon > 0$ *such that* $\langle L, x \rangle + \varepsilon \leq 0$ *for each* $x \in C$.

Clearly, Theorem 1.11 is a special case of Theorem 1.10. To verify that they are actually equivalent, simply write $C = A - B$ and observe that C is open if A is, while C is closed if A is compact and B is closed.

Second, part ii) of Theorem 1.11 can be easily deduced from part i) of Theorem 1.10 by considering a sufficiently small open ball A around the origin (not intersecting C), and writing $B = C$.

Finally, in finite-dimensional spaces, part i) of Theorem 1.11 can be obtained without any topological assumptions on the sets involved. More precisely, we have the following:

Proposition 1.12. *Given* $N \geq 1$, *let C be a nonempty and convex subset of* $\mathbf{R}^N$ *not containing the origin. Then, there exists* $v \in \mathbf{R}^N \setminus \{0\}$ *such that* $v \cdot x \leq 0$ *for each* $x \in C$. *In particular, if* $N \geq 2$ *and C is open, then*

$$V = \{x \in \mathbf{R}^N : v \cdot x = 0\}$$

is a nontrivial subspace of $\mathbf{R}^N$ *that does not intersect C.*

Proof. Let $(x_n) \in C$ such that the set $\{x_n : n \geq 1\}$ is dense in C. Let C_n be the convex hull of the set $\{x_k : k = 1, \ldots, n\}$ and let p_n be the least-norm element of C_n. By convexity, for each $x \in C_n$ and $t \in (0,1)$, we have

$$\|p_n\|^2 \leq \|p_n + t(x - p_n)\|^2 = \|p_n\|^2 + 2t\, p_n \cdot (x - p_n) + t^2 \|x - p_n\|^2.$$

Therefore,

$$0 \leq 2\|p_n\|^2 \leq 2\, p_n \cdot x + t\|x - p_n\|^2.$$

Letting $t \to 0$, we deduce that $p_n \cdot x \geq 0$ for all $x \in C_n$. Now write $v_n = -p_n/\|p_n\|$. The sequence (v_n) lies in the unit sphere, which is compact. We may extract a subsequence that converges to some $v \in \mathbf{R}^N$ with $\|v\| = 1$ (thus $v \neq 0$) and $v \cdot x \leq 0$ for all $x \in C$. $\square$

Proof of Theorem 1.11

Many standard functional analysis textbooks begin by presenting a general form of the Hahn–Banach Extension Theorem (see Theorem 1.13 below) and used to prove Theorem 1.11. We preferred to take the opposite path here, which has a more *convex-analytic* flavor. The same approach can be found, for instance, in [96].

Step 1: If the dimension of X is at least 2, there is a nontrivial subspace of X not intersecting C.

Take any two-dimensional subspace Y of X. If $Y \cap C = \emptyset$ there is nothing to prove. Otherwise, identify Y with $\mathbf{R}^2$ and use Proposition 1.12 to obtain a subspace of Y disjoint from $Y \cap C$, which clearly gives a subspace of X not intersecting C.

Step 2: There is a closed subspace M of X such that $M \cap C = \emptyset$ and the quotient space X/M has dimension 1.

Let $\mathscr{M}$ be the collection of all subspaces of X not intersecting C, ordered by inclusion. Step 1 shows that $\mathscr{M} \neq \emptyset$. According to Zorn's Lemma (see, for instance, [30, Lemma 1.1]), $\mathscr{M}$ has a maximal element M, which must be a closed subspace of X not intersecting C. The dimension of the quotient space X/M is at least 1 because $M \neq X$. The canonical homomorphism $\pi : X \to X/M$ is continuous and open. If the dimension of X/M is greater than 1, we use Step 1 again with $\tilde{X} = X/M$ and $\tilde{C} = \pi(C)$ to find a nontrivial subspace $\tilde{M}$ of $\tilde{X}$ that does not intersect $\tilde{C}$. Then $\pi^{-1}(\tilde{M})$ is a subspace of X that does not intersect C and is strictly greater than M, contradicting the maximality of the latter.

Step 3: Conclusion.

Take any (necessarily continuous) isomorphism $\phi : X/M \to \mathbf{R}$ and set $L = \phi \circ \pi$. Then, either $\langle L, x \rangle < 0$ for all $x \in C$, or $\langle -L, x \rangle < 0$ for all $x \in C$.

A Few Direct but Important Consequences

The following is known as the Hahn–Banach Extension Theorem:

Theorem 1.13. *Let M be a subspace of X and let $\ell : M \to \mathbf{R}$ be a linear function such that $\langle \ell, x \rangle \leq \alpha \|x\|$ for some $\alpha > 0$ and all $x \in M$. Then, there exists $L \in X^*$ such that $\langle L, x \rangle = \langle \ell, x \rangle$ for all $x \in M$ and $\|L\|_* \leq \alpha$.*

Proof. Define

$$A = \{ (x, \mu) \in X \times \mathbf{R} : \mu > \alpha \|x\|, x \in X \}$$
$$B = \{ (y, v) \in X \times \mathbf{R} : v = \langle \ell, y \rangle, y \in M \}.$$

By the Hahn–Banach Separation Theorem 1.10, there is $(\tilde{L}, s) \in X^* \times \mathbf{R} \setminus \{(0,0)\}$ such that

$$\langle \tilde{L}, x \rangle + s\mu \leq \langle \tilde{L}, y \rangle + sv$$

for all $(x, \mu) \in A$ and $(y, v) \in B$. Taking $x = y = 0$, $\mu = 1$ and $v = 0$, we deduce that $s \leq 0$. If $s = 0$, then $\langle \tilde{L}, x - y \rangle \leq 0$ for all $x \in X$ and so $\tilde{L} = 0$, which contradicts the fact that $(\tilde{L}, s) \neq (0,0)$. We conclude that $s > 0$. Writing $L = -\tilde{L}/s$, we obtain

$$\langle L, x \rangle - \mu \leq \langle L - \ell, y \rangle$$

for all $(x,\mu) \in A$ and $y \in M$. Passing to the limit as $\mu \to \alpha\|x\|$, we get

$$\langle L,x\rangle \leq \langle L-\ell,y\rangle + \alpha\|x\|$$

for all $x \in X$ and all $y \in M$. It follows that $L = \ell$ on M and $\|L\|_* \leq \alpha$. $\quad\square$

Another consequence of the Hahn–Banach Separation Theorem 1.10 is the following:

Corollary 1.14. *For each $x \in X$, there exists $\ell_x \in X^*$ such that $\|\ell_x\|_* = 1$ and $\langle \ell_x,x\rangle = \|x\|$.*

Proof. Set $A = B(0,\|x\|)$ and $B = \{x\}$. By Theorem 1.10, there exists $L_r \in X \setminus \{0\}$ such that $\langle L_x,y\rangle \leq \langle L_x,x\rangle$ for all $y \in A$. This implies $\langle L_x,x\rangle = \|L_x\|_* \|x\|$. The functional $\ell_x = L_x/\|L_x\|_*$ has the desired properties. $\quad\square$

The functional ℓ_x, given by Corollary 1.14, is a *support functional* for x. The *normalized duality mapping* is the set-valued function $\mathscr{F}: X \to \mathscr{P}(X^*)$ given by

$$\mathscr{F}(x) = \{x^* \in X^* : \|x^*\|_* = 1 \text{ and } \langle x^*,x\rangle = \|x\|\}.$$

The set $\mathscr{F}(x)$ is always convex and it need not be a singleton, as shown in the following example:

Example 1.15. Consider $X = \mathbf{R}^2$ with $\|(x_1,x_2)\| = |x_1| + |x_2|$ for $(x_1,x_2) \in X$. Then, $X^* = \mathbf{R}^2$ with $\langle(x_1^*,x_2^*),(x_1,x_2)\rangle = x_1^*x_1 + x_2^*x_2$ and $\|(x_1^*,x_2^*)\|_* = \max\{|x_1^*|,|x_2^*|\}$ for $(x_1^*,x_2^*) \in X^*$. Moreover, $\mathscr{F}(1,0) = \{(1,b) \in \mathbf{R}^2 : b \in [-1,1]\}$. $\quad\square$

From Corollary 1.14 we deduce the following:

Corollary 1.16. *For every $x \in X$, $\|x\| = \max_{\|L\|_*=1} \langle L,x\rangle$.*

Recall from Sect. 1.1.1 that the canonical embedding $\mathscr{J}$ of X into X^{**} is defined by $\mathscr{J}(x) = \mu_x$, where μ_x satisfies

$$\langle\mu_x,L\rangle_{X^{**},X^*} = \langle L,x\rangle_{X^*,X}.$$

Recall also that $\mathscr{J}$ is linear, and $\|\mathscr{J}(x)\|_{**} \leq \|x\|$ for all $x \in X$. We have the following:

Proposition 1.17. *The canonical embedding $\mathscr{J}: X \to X^{**}$ is a linear isometry.*

Proof. It remains to prove that $\|x\| \leq \|\mu_x\|_{**}$. To this end, simply notice that

$$\|\mu_x\|_{**} = \sup_{\|L\|_*=1} \mu_x(L) \geq \mu_x(\ell_x) = \langle \ell_x,x\rangle_{X^*,X} = \|x\|,$$

where ℓ_x is the functional given by Corollary 1.14. $\quad\square$

Another consequence of Corollary 1.16 concerns the adjoint of a bounded linear operator, defined in Sect. 1.1.1:

Corollary 1.18. *Let* $(X, \|\cdot\|_X)$ *and* $(Y, \|\cdot\|_Y)$ *be normed spaces. Given* $L \in \mathscr{L}(X;Y)$, *let* $L^* \in \mathscr{L}(Y^*;X^*)$ *be its adjoint. Then* $\|L^*\|_{\mathscr{L}(Y^*;X^*)} = \|L\|_{\mathscr{L}(X;Y)}$.

Proof. We already proved that $\|L^*\|_{\mathscr{L}(Y^*;X^*)} \le \|L\|_{\mathscr{L}(X;Y)}$. For the reverse inequality, use Corollary 1.16 to deduce that

$$\|L\|_{\mathscr{L}(X;Y)} = \sup_{\|x\|_X=1} \left[\max_{\|y^*\|_{Y^*}=1} \langle L^*(y^*), x \rangle_{X^*,X} \right] \le \|L^*\|_{\mathscr{L}(Y^*;X^*)},$$

which gives the result.   □

1.1.3 The Weak Topology

By definition, each element of X^* is continuous as a function from $(X, \|\cdot\|)$ to $(\mathbf{R}, |\cdot|)$. However, there are other topologies on X for which every element of X^* is continuous.[1] The coarsest of such topologies (the one with the fewest open sets) is called the *weak topology* and will be denoted by $\sigma(X)$, or simply σ, if there is no possible confusion.

Given $x_0 \in X$, $L \in X^*$ and $\varepsilon > 0$, every set of the form

$$\mathscr{V}_L^\varepsilon(x_0) = \{ x \in X \,:\, \langle L, x - x_0 \rangle < \varepsilon \} = L^{-1}\left((-\infty, L(x_0) + \varepsilon) \right)$$

is open for the weak topology and contains x_0. Moreover, the collection of all such sets generates a base of neighborhoods of x_0 for the weak topology in the sense that if V_0 is a neighborhood of x_0, then there exist $L_1, \dots, L_N \in X^*$ and $\varepsilon > 0$ such that

$$x_0 \in \bigcap_{k=1}^N \mathscr{V}_{L_k}^\varepsilon(x_0) \subset V_0.$$

Recall that a *Hausdorff space* is a topological space in which every two distinct points admit disjoint neighborhoods.

Proposition 1.19. (X, σ) *is a Hausdorff space.*

Proof. Let $x_1 \ne x_2$. Part ii) of the Hahn–Banach Separation Theorem 1.10 shows the existence of $L \in X^*$ such that $\langle L, x_1 \rangle + \varepsilon \le \langle L, x_2 \rangle$. Then $\mathscr{V}_L^{\frac{\varepsilon}{2}}(x_1)$ and $\mathscr{V}_{-L}^{\frac{\varepsilon}{2}}(x_2)$ are disjoint weakly open sets containing x_1 and x_2, respectively.   □

It follows from the definition that every weakly open set is (strongly) open. If X is finite-dimensional, the weak topology coincides with the strong topology. Roughly

---

[1] A trivial example is the *discrete topology* (for which every set is open), but it is not very useful for our purposes.

speaking, the main idea is that, inside any open ball, one can find a weak neighborhood of its center that can be expressed as a *finite* intersection of open half-spaces.

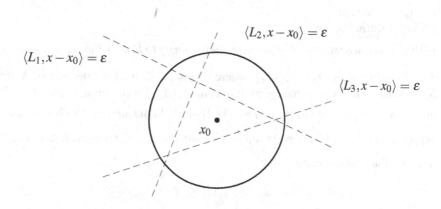

On the other hand, if X is infinite-dimensional, the weak topology is strictly coarser than the strong topology, as shown in the following example:

Example 1.20. If X is infinite-dimensional, the ball $B(0,1)$ is open for the strong topology but not for the weak topology. Roughly speaking, the reason is that no *finite* intersection of half-spaces can be bounded in all directions.   □

In other words, there are open sets which are not weakly open. Of course, the same is true for closed sets. However, closed convex sets are weakly closed.

Proposition 1.21. *A convex subset of a normed space is closed for the strong topology if, and only if, it is closed for the weak topology.*

Proof. By definition, every weakly closed set must be strongly closed. Conversely, let $C \subset X$ be convex and strongly closed. Given $x_0 \notin C$, we may apply part ii) of the Hahn–Banach Separation Theorem 1.10 with $A = \{x_0\}$ and $B = C$ to deduce the existence of $L \in X^* \setminus \{0\}$ and $\varepsilon > 0$ such that $\langle L, x_0 \rangle + \varepsilon \leq \langle L, y \rangle$ for each $y \in C$. The set $\mathscr{V}_L^\varepsilon(x_0)$ is a weak neighborhood of x_0 that does not intersect C. It follows that C^c is weakly open.   □

Weakly Convergent Sequences

We say that a sequence (x_n) in X *converges weakly* to $\bar{x}$, and write $x_n \rightharpoonup \bar{x}$, as $n \to \infty$ if $\lim_{n \to \infty} \langle L, x_n - \bar{x} \rangle = 0$ for each $L \in X^*$. This is equivalent to saying that for each weakly open neighborhood $\mathscr{V}$ of $\bar{x}$ there is $N \in \mathbf{N}$ such that $x_n \in \mathscr{V}$ for all $n \geq N$. The point $\bar{x}$ is the *weak limit* of the sequence.

Since $|\langle L, x_n - \bar{x}\rangle| \leq \|L\|_* \|x_n - \bar{x}\|$, strongly convergent sequences are weakly convergent and the limits coincide.

Proposition 1.22. *Let* (x_n) *converge weakly to* $\bar{x}$ *as* $n \to \infty$*. Then:*

i) (x_n) *is bounded.*

ii) $\|\bar{x}\| \leq \liminf\limits_{n\to\infty} \|x_n\|$.

iii) *If* (L_n) *is a sequence in* X^* *that converges strongly to* $\bar{L}$*, then* $\lim\limits_{n\to\infty}\langle L_n, x_n\rangle = \langle \bar{L}, \bar{x}\rangle$.

Proof. For i), write $\mu_n = \mathscr{J}(x_n)$, where $\mathscr{J}$ is the canonical injection of X into X^{**}, which is a linear isometry, by Proposition 1.17. Since $\lim\limits_{n\to\infty} \mu_n(L) = \langle L, \bar{x}\rangle$ for all $L \in X^*$, we have $\sup\limits_{n\in\mathbf{N}} \mu_n(L) < +\infty$. The Banach–Steinhaus uniform boundedness principle (Theorem 1.9) implies $\sup\limits_{n\in\mathbf{N}} \|x_n\| = \sup\limits_{n\in\mathbf{N}} \|\mu_n\|_{**} \leq C$ for some $C > 0$. For ii), use Corollary 1.14 to deduce that

$$\|x\| = \langle \ell_{\bar{x}}, x - x_n\rangle + \langle \ell_{\bar{x}}, x_n\rangle \leq \langle \ell_{\bar{x}}, x - x_n\rangle + \|x_n\|,$$

and let $n \to \infty$. Finally, by part i), we have

$$\begin{aligned}|\langle L_n, x_n\rangle - \langle \bar{L}, \bar{x}\rangle| &\leq |\langle L_n - \bar{L}, x_n\rangle| + |\langle \bar{L}, x_n - \bar{x}\rangle| \\ &\leq C\|L_n - \bar{L}\|_* + |\langle \bar{L}, x_n - \bar{x}\rangle|.\end{aligned}$$

As $n \to \infty$, we obtain iii). $\qquad\square$

More on Closed Sets

Since we have defined two topologies on X, there exist (strongly) closed sets and weakly closed sets. It is possible and very useful to define some *sequential* notions as well. A set $C \subset X$ is *sequentially closed* if every convergent sequence of points in C has its limit in C. Analogously, C is *weakly sequentially closed* if every weakly convergent sequence in C has its weak limit in C. The relationship between the various notions of closedness is summarized in the following result:

Proposition 1.23. *Consider the following statements concerning a nonempty set* $C \subset X$:

i) C *is weakly closed.*

ii) C *is weakly sequentially closed.*

iii) C *is sequentially closed.*

iv) C *is closed.*

Then i) $\Rightarrow$ ii) $\Rightarrow$ iii) $\Leftrightarrow$ iv) $\Leftarrow$ i). *If* C *is convex, the four statements are equivalent.*

Proof. It is easy to show that i) $\Rightarrow$ ii) and iii) $\Leftrightarrow$ iv). We also know that i) $\Rightarrow$ iv) and ii) $\Rightarrow$ iii) because the weak topology is coarser than the strong topology. Finally, if C is convex, Proposition 1.21 states precisely that i) $\Leftrightarrow$ iv), which closes the loop. $\quad\square$

The Topological Dual Revisited: The Weak* Topology

The topological dual X^* of a normed space $(X, \|\cdot\|)$ is a Banach space with the norm $\|\cdot\|_*$. As in Sect. 1.1.3, we can define the weak topology $\sigma(X^*)$ in X^*. Recall that a base of neighborhoods for some point $L \in X^*$ is generated by the sets of the form

$$\{\ell \in X^* : \langle \mu, \ell - L \rangle_{X^{**},X^*} < \varepsilon\}, \tag{1.1}$$

with $\mu \in X^{**}$ and $\varepsilon > 0$.

Nevertheless, since X^* is, by definition, a space of functions, a third topology can be defined on X^* in a very natural way. It is the *topology of pointwise convergence*, which is usually referred to as the *weak* topology* in this context. We shall denote it by $\sigma^*(X^*)$, or simply σ^* if the space is clear from the context. For this topology, a base of neighborhoods for a point $L \in X^*$ is generated by the sets of the form

$$\{\ell \in X^* : \langle \ell - L, x \rangle_{X^*,X} < \varepsilon\}, \tag{1.2}$$

with $x \in X$ and $\varepsilon > 0$. Notice the similarity and difference with (1.1). Now, since every $x \in X$ determines an element $\mu_x \in X^{**}$ by the relation

$$\langle \mu_x, \ell \rangle_{X^{**},X^*} = \langle \ell, x \rangle_{X^*,X},$$

it is clear that every set that is open for the weak* topology must be open for the weak topology as well. In other words, $\sigma^* \subset \sigma$.

Reflexivity and Weak Compactness

In infinite-dimensional normed spaces, compact sets are rather scarce. For instance, in such spaces the closed balls are *not* compact (see [30, Theorem 6.5]). One of the most important properties of the weak* topology is that, according to the Banach–Alaoglu Theorem (see, for instance, [30, Theorem 3.16]), the closed unit ball in X^* is compact for the weak* topology. Recall, from Sect. 1.1.1, that X is reflexive if the canonical embedding $\mathscr{J}$ of X into X^{**} is surjective. This implies that the spaces (X, σ) and (X^{**}, σ^*) are homeomorphic, and so, the closed unit ball of X is compact for the weak topology. Further, we have the following:

Theorem 1.24. *Let $(X, \|\cdot\|)$ be a Banach space. The following are equivalent:*

i) X is reflexive.
ii) The closed unit ball $\bar{B}(0, 1)$ is compact for the weak topology.
iii) Every bounded sequence in X has a weakly convergent subsequence.

We shall not include the proof here for the sake of brevity. The interested reader may consult [50, Chaps. II and V] for full detail, or [30, Chap. 3] for abridged commentaries.

An important consequence is the following:

Corollary 1.25. *Let (y_n) be a bounded sequence in a reflexive space. If every weakly convergent subsequence has the same weak limit $\hat{y}$, then (y_n) must converge weakly to $\hat{y}$ as $n \to \infty$.*

Proof. Suppose (y_n) does not converge weakly to $\hat{y}$. Then, there exist a weakly open neighborhood $\mathscr{V}$ of $\hat{y}$, and a subsequence (y_{k_n}) of (y_n) such that $y_{k_n} \notin \mathscr{V}$ for all $n \in \mathbf{N}$. Since (y_{k_n}) is bounded, it has a subsequence $(y_{j_{k_n}})$ that converges weakly as $n \to \infty$ to some $\check{y}$ which cannot be in $\mathscr{V}$ and so $\check{y} \neq \hat{y}$. This contradicts the uniqueness of $\hat{y}$. $\qquad\square$

1.1.4 Differential Calculus

Consider a nonempty open set $A \subset X$ and function $f : A \to \mathbf{R}$. The *directional derivative* of f at $x \in A$ in the direction $d \in X$ is

$$f'(x;d) = \lim_{t \to 0^+} \frac{f(x+td) - f(x)}{t},$$

whenever this limit exists. The function f is *differentiable* (or simply *Gâteux-differentiable*) at x if $f'(x;d)$ exists for all $d \in X$ and the function $d \mapsto f'(x;d)$ is linear and continuous. In this situation, the *Gâteaux derivative* (or *gradient*) of f at x is $\nabla f(x) = f'(x;\cdot)$, which is an element of X^*. On the other hand, f is *differentiable in the sense of Fréchet* (or *Fréchet-differentiable*) at x if there exists $L \in X^*$ such that

$$\lim_{\|h\| \to 0} \frac{|f(x+h) - f(x) - \langle L, h \rangle|}{\|h\|} = 0.$$

If it is the case, the *Fréchet derivative* of f at x is $Df(x) = L$. An immediate consequence of these definitions is

Proposition 1.26. *If f is Fréchet-differentiable at x, then it is continuous and Gâteaux-differentiable there, with $\nabla f(x) = Df(x)$.*

As usual, f is differentiable (in the sense of Gâteaux or Fréchet) on A if it is so at every point of A.

Example 1.27. Let $B : X \times X \to \mathbf{R}$ be a bilinear function:

$$B(x+\alpha y, z) = B(x,z) + \alpha B(y,z) \quad \text{and} \quad B(x, y+\alpha z) = B(x,y) + \alpha B(x,z)$$

for all $x, y, z \in X$ and $\alpha \in \mathbf{R}$. Suppose also that B is continuous: $|B(x,y)| \leq \beta \|x\| \|y\|$ for some $\beta \geq 0$ and all $x, y \in X$. The function $f : X \to \mathbf{R}$, defined by $f(x) = B(x,x)$, is Fréchet-differentiable and $Df(x)h = B(x,h) + B(h,x)$. Of course, if B is symmetric: $B(x,y) = B(y,x)$ for all $x, y \in X$, then $Df(x)h = 2B(x,h)$. $\qquad\square$

Example 1.28. Let X be the space of continuously differentiable functions defined on $[0,T]$ with values in $\mathbf{R}^N$, equipped with the norm

$$\|x\|_X = \max_{t \in [0,T]} \|x(t)\| + \max_{t \in [0,T]} \|\dot{x}(t)\|.$$

Given a continuously differentiable function $\ell : \mathbf{R} \times \mathbf{R}^N \times \mathbf{R}^N \to \mathbf{R}$, define $J : X \to \mathbf{R}$ by

$$J[u] = \int_0^T \ell(t, x(t), \dot{x}(t)) \, dt.$$

Then J is Fréchet-differentiable and

$$DJ(x)h = \int_0^T \left[\nabla_2 \ell(t, x(t), \dot{x}(t)) \cdot h(t) + \nabla_3 \ell(t, x(t), \dot{x}(t)) \cdot \dot{h}(t) \right] dt,$$

where we use ∇_i to denote the gradient with respect to the i-th set of variables. $\square$

It is to note that the Gâteaux-differentiability does not imply continuity. In particular, it is weaker than Fréchet-differentiability.

Example 1.29. Define $f : \mathbf{R}^2 \to \mathbf{R}$ by

$$f(x,y) = \begin{cases} \dfrac{2x^4 y}{x^6 + y^3} & \text{if } (x,y) \neq (0,0) \\ 0 & \text{if } (x,y) = (0,0). \end{cases}$$

A simple computation shows that $\nabla f(0,0) = (0,0)$. However, $\lim_{z \to 0} f(z,z^2) = 1 \neq f(0,0)$. $\square$

If the gradient of f is Lipschitz-continuous, we can obtain a more precise first-order estimation for the values of the function:

Lemma 1.30 (Descent Lemma). *If $f : X \to \mathbf{R}$ is Gâteaux-differentiable and ∇f is Lipschitz-continuous with constant L, then*

$$f(y) \leq f(x) + \langle \nabla f(x), y - x \rangle + \frac{L}{2} \|y - x\|^2$$

for each $x, y \in X$. In particular, f is continuous.

Proof. Write $h = y - x$ and define $g : [0,1] \to \mathbf{R}$ by $g(t) = f(x + th)$. Then $\dot{g}(t) = \langle \nabla f(x + th), h \rangle$ for each $t \in (0,1)$, and so

$$\int_0^1 \langle \nabla f(x + th), h \rangle \, dt = \int_0^1 \dot{g}(t) \, dt = g(1) - g(0) = f(y) - f(x).$$

Therefore,

$$f(y) - f(x) = \int_0^1 \langle \nabla f(x), h \rangle \, dt + \int_0^1 \langle \nabla f(x+th) - \nabla f(x), h \rangle \, dt$$

$$\leq \langle \nabla f(x), h \rangle + \int_0^1 \| \nabla f(x+th) - \nabla f(x) \| \, \| h \| \, dt$$

$$\leq \langle \nabla f(x), h \rangle + L \| h \|^2 \int_0^1 t \, dt$$

$$= \langle \nabla f(x), y - x \rangle + \frac{L}{2} \| y - x \|^2,$$

as claimed. $\square$

Second Derivatives

If $f : A \to \mathbf{R}$ is Gâteaux-differentiable in A, a valid question is whether the function $\nabla f : A \to X^*$ is differentiable. As before, one can define a directional derivative

$$(\nabla f)'(x; d) = \lim_{t \to 0^+} \frac{\nabla f(x+td) - \nabla f(x)}{t},$$

whenever this limit exists (with respect to the strong topology of X^*). The function f is *twice differentiable in the sense of Gâteaux* (or simply *twice Gâteaux-differentiable*) in x if f is Gâteaux-differentiable in a neighborhood of x, $(\nabla f)'(x; d)$ exists for all $d \in X$, and the function $d \mapsto (\nabla f)'(x; d)$ is linear and continuous. In this situation, the *second Gâteaux derivative* (or *Hessian*) of f at x is $\nabla^2 f(x) = (\nabla f)'(x; \cdot)$, which is an element of $\mathscr{L}(X; X^*)$. Similarly, f is *twice differentiable in the sense of Fréchet* (or *twice Fréchet-differentiable*) at x if there exists $M \in \mathscr{L}(X; X^*)$ such that

$$\lim_{\|h\| \to 0} \frac{\| Df(x+h) - Df(x) - M(h) \|_*}{\|h\|} = 0.$$

The *second Fréchet derivative* of f at x is $D^2 f(x) = M$.

We have the following:

Proposition 1.31 (Second-order Taylor Approximation). *Let A be an open subset of X and let $x \in A$. Assume $f : A \to \mathbf{R}$ is twice Gâteaux-differentiable in x. Then, for each $d \in X$, we have*

$$\lim_{t \to 0} \frac{1}{t^2} \left| f(x+td) - f(x) - t \langle \nabla f(x), d \rangle - \frac{t^2}{2} \langle \nabla^2 f(x) d, d \rangle \right| = 0.$$

Proof. Define $\phi : I \subset R \to \mathbf{R}$ by $\phi(t) = f(x+td)$, where I is a sufficiently small open interval around 0 such that $\phi(t)$ exists for all $t \in I$. It is easy to see that $\phi'(t) =$

$\langle \nabla f(x+td), d \rangle$ and $\phi''(0) = \langle \nabla^2 f(x)d, d \rangle$. The second-order Taylor expansion for ϕ in $\mathbf{R}$ yields

$$\lim_{t \to 0} \frac{1}{t^2} \left| \phi(t) - \phi(0) - t\phi'(0) - \frac{t^2}{2}\phi''(0) \right| = 0,$$

which gives the result. □

Of course, it is possible to define derivatives of higher order, and obtain the corresponding Taylor approximations.

Optimality Conditions for Differentiable Optimization Problems

The following is the keynote necessary condition for a point $\hat{x}$ to minimize a Gâteaux-differentiable function f over a convex set C.

Theorem 1.32 (Fermat's Rule). *Let C be a convex subset of a normed space $(X, \|\cdot\|)$ and let $f : X \to \mathbf{R} \cup \{+\infty\}$. If $f(\hat{x}) \leq f(y)$ for all $y \in C$ and if f is Gâteaux-differentiable at $\hat{x}$, then*

$$\langle \nabla f(\hat{x}), y - \hat{x} \rangle \geq 0$$

for all $y \in C$. If moreover $\hat{x} \in \mathrm{int}(C)$, then $\nabla f(\hat{x}) = 0$.

Proof. Take $y \in C$. Since C is convex, for each $\lambda \in (0,1)$, the point $y_\lambda = \lambda y + (1 - \lambda)\hat{x}$ belongs to C. The inequality $f(\hat{x}) \leq f(y_\lambda)$ is equivalent to $f(\hat{x} + \lambda(y - \hat{x})) - f(\hat{x}) \geq 0$. It suffices to divide by λ and let $\lambda \to 0$ to deduce that $f'(\hat{x}; y - \hat{x}) \geq 0$ for all $y \in C$. □

To fix the ideas, consider a differentiable function on $X = \mathbf{R}^2$. Theorem 1.32 states that the gradient of f at $\hat{x}$ must point *inwards*, with respect to C. In other words, f can only decrease by leaving the set C. This situation is depicted below:

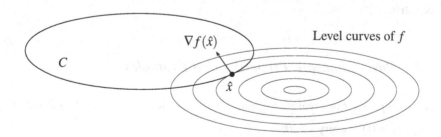

As we shall see in the next chapter, the condition given by Fermat's Rule (Theorem 1.32) is not only necessary, but also sufficient, for convex functions. In the general case, one can provide second-order necessary and sufficient conditions for optimality. We state this result in the unconstrained case $(C = X)$ for simplicity.

An operator $M \in \mathscr{L}(X;X^*)$ is *positive semidefinite* if $\langle Md,d \rangle \geq 0$ for all $d \in X$; *positive definite* if $\langle Md,d \rangle > 0$ for all $d \neq 0$; and *uniformly elliptic* with constant $\alpha > 0$ if $\langle Md,d \rangle \geq \frac{\alpha}{2}\|d\|^2$ for all $d \in X$.

Theorem 1.33. *Let A be an open subset of a normed space $(X, \|\cdot\|)$, let $\hat{x} \in A$, and let $f : A \to \mathbf{R}$.*

i) *If $f(\hat{x}) \leq f(y)$ for all y in a neighborhood of $\hat{x}$, and f is twice Gâteaux-differentiable at $\hat{x}$, then $\nabla f(\hat{x}) = 0$ and $\nabla^2 f(\hat{x})$ is positive semidefinite.*

ii) *If $\nabla f(\hat{x}) = 0$ and $\nabla^2 f(\hat{x})$ is uniformly elliptic, then $f(\hat{x}) < f(y)$ for all y in a neighborhood of $\hat{x}$.*

Proof. For i), we already know by Theorem 1.32 that $\nabla f(\hat{x}) = 0$. Now, if $d \in X$ and $\varepsilon > 0$, by Proposition 1.31, there is $t_0 > 0$ such that

$$\frac{t^2}{2}\langle \nabla^2 f(\hat{x})d,d \rangle > f(\hat{x}+td) - f(\hat{x}) - \varepsilon t^2 \geq -\varepsilon t^2$$

for all $t \in [0,t_0]$. It follows that $\langle \nabla^2 f(\hat{x})d,d \rangle \geq 0$.

For ii), assume $\nabla^2 f(\hat{x})$ is uniformly elliptic with constant $\alpha > 0$ and take $d \in X$. Set $\varepsilon = \frac{\alpha}{4}\|d\|^2$. By Proposition 1.31, there is $t_1 > 0$ such that

$$f(\hat{x}+td) > f(\hat{x}) + \frac{t^2}{2}\langle \nabla^2 f(\hat{x})d,d \rangle - \varepsilon t^2 \geq f(\hat{x})$$

for all $t \in [0,t_1]$. $\square$

1.2 Hilbert Spaces

Hilbert spaces are an important class of Banach spaces with rich geometric properties.

1.2.1 Basic Concepts, Properties and Examples

An *inner product* in a real vector space H is a function $\langle \cdot, \cdot \rangle : H \times H \to \mathbf{R}$ such that:

i) $\langle x,x \rangle > 0$ for every $x \neq 0$;

ii) $\langle x,y \rangle = \langle y,x \rangle$ for each $x,y \in H$;

iii) $\langle \alpha x+y,z \rangle = \alpha\langle x,z \rangle + \langle y,z \rangle$ for each $\alpha \in \mathbf{R}$ and $x,y,z \in H$.

The function $\|\cdot\| : H \to \mathbf{R}$, defined by $\|x\| = \sqrt{\langle x,x \rangle}$, is a *norm* on H. Indeed, it is clear that $\|x\| > 0$ for every $x \neq 0$. Moreover, for each $\alpha \in \mathbf{R}$ and $x \in H$, we have $\|\alpha x\| = |\alpha|\|x\|$. It only remains to verify the triangle inequality. We have the following:

Proposition 1.34. *For each $x, y \in H$ we have*

i) *The Cauchy–Schwarz inequality:* $|\langle x, y \rangle| \le \|x\| \, \|y\|$.
ii) *The triangle inequality:* $\|x + y\| \le \|x\| + \|y\|$.

Proof. The Cauchy–Schwarz inequality is trivially satisfied if $y = 0$. If $y \ne 0$ and $\alpha > 0$, then

$$0 \le \|x \pm \alpha y\|^2 = \langle x \pm \alpha y, x \pm \alpha y \rangle = \|x\|^2 \pm 2\alpha \langle x, y \rangle + \alpha^2 \|y\|^2.$$

Therefore,

$$|\langle x, y \rangle| \le \frac{1}{2\alpha} \|x\|^2 + \frac{\alpha}{2} \|y\|^2$$

for each $\alpha > 0$. In particular, taking $\alpha = \|x\| / \|y\|$, we deduce i). Next, we use i) to deduce that

$$\|x + y\|^2 = \|x\|^2 + 2\langle x, y \rangle + \|y\|^2 \le \|x\|^2 + 2\|x\| \, \|y\| + \|y\|^2 = (\|x\| + \|y\|)^2,$$

whence ii) holds. $\qquad\qquad\qquad\qquad\qquad\qquad\qquad\qquad\qquad\qquad\qquad\qquad\qquad$ $\square$

If $\|x\| = \sqrt{\langle x, x \rangle}$ for all $x \in X$, we say that the norm $\|\cdot\|$ is *associated* to the inner product $\langle \cdot, \cdot \rangle$. A *Hilbert space* is a Banach space, whose norm is associated to an inner product.

Example 1.35. The following are Hilbert spaces:

i) The Euclidean space $\mathbf{R}^N$ is a Hilbert space with the inner product given by the *dot product*: $\langle x, y \rangle = x \cdot y$.
ii) The space $\ell^2(\mathbf{N}; \mathbf{R})$ of real sequences $\mathbf{x} = (x_n)$ such that

$$\sum_{n \in \mathbf{N}} x_n^2 < +\infty,$$

equipped with the inner product defined by $\langle \mathbf{x}, \mathbf{y} \rangle = \sum_{n \in \mathbf{N}} x_n y_n$.
iii) Let Ω be a bounded open subset of $\mathbf{R}^N$. The space $L^2(\Omega; \mathbf{R}^M)$ of (classes of) measurable vector fields $\phi : \Omega \to \mathbf{R}^M$ such that

$$\int_\Omega \phi_m(\zeta)^2 \, d\zeta < +\infty, \qquad \text{for} \quad m = 1, 2, \ldots, M,$$

with the inner product $\langle \phi, \psi \rangle = \sum_{m=1}^M \int_\Omega \phi_m(\zeta) \, \psi_m(\zeta) \, d\zeta$. $\qquad\qquad\qquad$ $\square$

By analogy with $\mathbf{R}^N$, it seems reasonable to define the angle γ between two *nonzero* vectors x and y by the relation

$$\cos(\gamma) = \frac{\langle x, y \rangle}{\|x\| \, \|y\|}, \qquad \gamma \in [0, \pi].$$

We shall say that x and y are *orthogonal*, and write $x \perp y$, if $\cos(\gamma) = 0$. In a similar fashion, we say x and y are *parallel*, and write $x \| y$, if $|\cos(\gamma)| = 1$. With this notation, we have

i) *Pythagoras Theorem*: $x \perp y$ if, and only if, $\|x+y\|^2 = \|x\|^2 + \|y\|^2$;

ii) The *colinearity condition*: $x \,\|\, y$ if, and only if, $x = \lambda y$ with $\lambda \in \mathbf{R}$.

Another important geometric property of the norm in a Hilbert space is the *Parallelogram Identity*, which states that

$$\|x+y\|^2 + \|x-y\|^2 = 2\left(\|x\|^2 + \|y\|^2\right) \tag{1.3}$$

for each $x, y \in H$. It shows the relationship between the length of the sides and the lengths of the diagonals in a parallelogram, and is easily proved by adding the following identities

$$\|x+y\|^2 = \|x\|^2 + \|y\|^2 + 2\langle x,y \rangle \quad \text{and} \quad \|x-y\|^2 = \|x\|^2 + \|y\|^2 - 2\langle x,y \rangle.$$

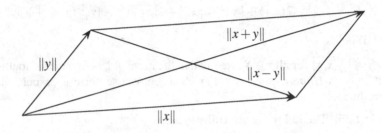

Example 1.36. The space $X = \mathscr{C}([0,1]; \mathbf{R})$ with the norm $\|x\| = \max\limits_{t \in [0,1]} |x(t)|$ is not a Hilbert space. Consider the functions $x, y \in X$, defined by $x(t) = 1$ and $y(t) = t$ for $t \in [0,1]$. We have $\|x\| = 1$, $\|y\| = 1$, $\|x+y\| = 2$ and $\|x-y\| = 1$. The parallelogram identity (1.3) does not hold. $\qquad\square$

1.2.2 Projection and Orthogonality

An important property of Hilbert spaces is that given a nonempty, closed, and convex subset K of H and a point $x \notin K$, there is a unique point in K which is the closest to x. More precisely, we have the following:

Proposition 1.37. *Let K be a nonempty, closed, and convex subset of H and let $x \in H$. Then, there exists a unique point $y^* \in K$ such that*

$$\|x - y^*\| = \min\limits_{y \in K} \|x - y\|. \tag{1.4}$$

Moreover, it is the only element of K such that

$$\langle x - y^*, y - y^* \rangle \le 0 \quad \text{for all} \quad y \in K. \tag{1.5}$$

Proof. We shall prove Proposition 1.37 in three steps: first, we verify that (1.4) has a solution; next, we establish the equivalence between (1.4) and (1.5); and finally, we check that (1.5) cannot have more than one solution.

First, set $d = \inf_{y \in K} \|x - y\|$ and consider a sequence (y_n) in K such that $\lim_{n \to \infty} \|y_n - x\| = d$. We have

$$\|y_n - y_m\|^2 = \|(y_n - x) + (x - y_m)\|^2$$
$$= 2 \left(\|y_n - x\|^2 + \|y_m - x\|^2 \right) - \|(y_n + y_m) - 2x\|^2,$$

by virtue of the parallelogram identity (1.3). Since K is convex, the midpoint between y_n and y_m belongs to K. Therefore,

$$\|(y_n + y_m) - 2x\|^2 = 4 \left\| \frac{y_n + y_m}{2} - x \right\|^2 \geq 4d^2,$$

according to the definition of d. We deduce that

$$\|y_n - y_m\|^2 \leq 2 \left(\|y_n - x\|^2 + \|y_m - x\|^2 - 2d^2 \right).$$

Whence, (y_n) is a Cauchy sequence, and must converge to some y^*, which must lie in K by closedness. The continuity of the norm implies $d = \lim_{n \to \infty} \|y_n - x\| = \|y^* - x\|$.

Next, assume (1.4) holds and let $y \in K$. Since K is convex, for each $\lambda \in (0,1)$ the point $\lambda y + (1 - \lambda) y^*$ also belongs to K. Therefore,

$$\|x - y^*\|^2 \leq \|x - \lambda y - (1 - \lambda) y^*\|^2$$
$$= \|x - y^*\|^2 + 2\lambda(1 - \lambda)\langle x - y^*, y^* - y \rangle + \lambda^2 \|y^* - y\|^2.$$

This implies

$$\langle x - y^*, y - y^* \rangle \leq \frac{\lambda}{2(1 - \lambda)} \|y^* - y\|^2.$$

Letting $\lambda \to 0$ we obtain (1.5). Conversely, if (1.5) holds, then

$$\|x - y\|^2 = \|x - y^*\|^2 + 2\langle x - y^*, y^* - y \rangle + \|y^* - y\|^2 \geq \|x - y^*\|^2$$

for each $y \in K$ and (1.4) holds.

Finally, if $y_1^*, y_2^* \in K$ satisfy (1.5), then

$$\langle x - y_1^*, y_2^* - y_1^* \rangle \leq 0 \quad \text{and} \quad \langle x - y_2^*, y_1^* - y_2^* \rangle \leq 0.$$

Adding the two inequalities we deduce that $y_1^* = y_2^*$. $\square$

The point y^* given by Proposition 1.37 is the *projection* of x onto K and will be denoted by $P_K(x)$. The characterization of $P_K(x)$ given by (1.5) says that for each $x \notin K$, the set K lies in the *closed half-space*

$$S = \{\, y \in H : \langle x - P_K(x), y - P_K(x) \rangle \le 0 \,\}.$$

Corollary 1.38. *Let K be a nonempty, closed, and convex subset of H. Then $K = \cap_{x \notin K}\{y \in H : \langle x - P_K(x), y - P_K(x) \rangle\}$.*

Conversely, the intersection of closed convex half-spaces is closed and convex.

For subspaces we recover the idea of *orthogonal projection*:

Proposition 1.39. *Let M be a closed subspace of H. Then,*

$$\langle x - P_M(x), u \rangle = 0$$

for each $x \in H$ and $u \in M$. In other words, $x - P_M(x) \perp M$.

Proof. Let $u \in M$ and write $v_\pm = P_M(x) \pm u$. Then $v_\pm \in M$ and so

$$\pm \langle x - P_M(x), u \rangle \le 0.$$

It follows that $\langle x - P_M(x), u \rangle = 0$. □

Another property of the projection, bearing important topological consequences, is the following:

Proposition 1.40. *Let K be a nonempty, closed, and convex subset of H. The function $x \mapsto P_K(x)$ is nonexpansive.*

Proof. Let $x_1, x_2 \in H$. Then $\langle x_1 - P_K(x_1), P_K(x_2) - P_K(x_1) \rangle \le 0$ and $\langle x_2 - P_K(x_2), P_K(x_1) - P_K(x_2) \rangle \le 0$. Summing these two inequalities we obtain

$$\|P_K(x_1) - P_K(x_2)\|^2 \le \langle x_1 - x_2, P_K(x_1) - P_K(x_2) \rangle.$$

We conclude using the Cauchy–Schwarz inequality. □

1.2.3 Duality, Reflexivity and Weak Convergence

The topological dual of a real Hilbert space can be easily characterized. Given $y \in H$, the function $L_y : H \to \mathbf{R}$, defined by $L_y(h) = \langle y, h \rangle$, is linear and continuous by the Cauchy–Schwarz inequality. Moreover, $\|L_y\|_* = \|y\|$. Conversely, we have the following:

Theorem 1.41 (Riesz Representation Theorem). *Let $L : H \to \mathbf{R}$ be a continuous linear function on H. Then, there exists a unique $y_L \in H$ such that*

$$L(h) = \langle y_L, h \rangle$$

for each h ∈ H. Moreover, the function $L \mapsto y_L$ is a linear isometry.

Proof. Let $M = \ker(L)$, which is a closed subspace of H because L is linear and continuous. If $M = H$, then $L(h) = 0$ for all $h \in H$ and it suffices to take $y_L = 0$. If $M \neq H$, let us pick any $x \notin M$ and define

$$\hat{x} = x - P_M(x).$$

Notice that $\hat{x} \neq 0$ and $\hat{x} \notin M$. Given any $h \in H$, set $u_h = L(\hat{x})h - L(h)\hat{x}$, so that $u_h \in M$. By Proposition 1.39, we have $\langle \hat{x}, u_h \rangle = 0$. In other words,

$$0 = \langle \hat{x}, u_h \rangle = \langle \hat{x}, L(\hat{x})h - L(h)\hat{x} \rangle = L(\hat{x})\langle \hat{x}, h \rangle - L(h)\|\hat{x}\|^2.$$

The vector

$$y_L = \frac{L(\hat{x})}{\|\hat{x}\|^2}\hat{x}$$

has the desired property and it is straightforward to verify that the function $L \mapsto y_L$ is a linear isometry. $\square$

As a consequence, the inner product $\langle \cdot, \cdot \rangle_* : H^* \times H^*$ defined by

$$\langle L_1, L_2 \rangle_* = L_1(y_{L_2}) = \langle y_{L_1}, y_{L_2} \rangle$$

turns H^* into a Hilbert space, which is isometrically isomorphic to H. The norm associated with $\langle \cdot, \cdot \rangle_*$ is precisely $\| \cdot \|_*$.

Corollary 1.42. *Hilbert spaces are reflexive.*

Proof. Given $\mu \in H^{**}$, use the Riesz Representation Theorem 1.41 twice to obtain $L_\mu \in H^*$ such that $\mu(\ell) = \langle L_\mu, \ell \rangle_*$ for each $\ell \in H^*$, and then $y_{L_\mu} \in H$ such that $L_\mu(x) = \langle y_{L_\mu}, x \rangle$ for all $x \in H$. It follows that $\mu(\ell) = \langle L_\mu, \ell \rangle_* = \ell(y_{L_\mu})$ for each $\ell \in H^*$ by the definition of $\langle \cdot, \cdot \rangle_*$. $\square$

Remark 1.43. Theorem 1.41 also implies that a sequence (x_n) on a Hilbert space H converges weakly to some $x \in H$ if, and only if, $\lim_{n \to \infty} \langle x_n - x, y \rangle = 0$ for all $y \in H$. $\square$

Strong and weak convergence can be related as follows:

Proposition 1.44. *A sequence in (x_n) converges strongly to $\bar{x}$ if, and only if, it converges weakly to $\bar{x}$ and $\limsup_{n \to \infty} \|x_n\| \leq \|\bar{x}\|$.*

Proof. The *only if* part is immediate. For the *if* part, notice that $0 \leq \limsup_{n \to \infty} \|x_n - \bar{x}\|^2 = \limsup_{n \to \infty} \left[\|x_n\|^2 + \|\bar{x}\|^2 - 2\langle x_n, \bar{x} \rangle \right] \leq 0.$ $\square$

Remark 1.45. Another consequence of Theorem 1.41 is that we may interpret the gradient of a Gâteaux-differentiable function f as an element of H instead of H^* (see Sect. 1.1.4). This will be useful in the design of optimization algorithms (see Chap. 6). $\square$

Chapter 2
Existence of Minimizers

Abstract In this chapter, we present sufficient conditions for an extended real-valued function to have minimizers. After discussing the main concepts, we begin by addressing the existing issue in abstract Hausdorff spaces, under certain (one-sided) continuity and compactness hypotheses. We also present Ekeland's Variational Principle, providing the existence of approximate minimizers that are strict in some sense. Afterward, we study the minimization of convex functions in reflexive spaces, where the verification of the hypothesis is more practical. Although it is possible to focus directly on this setting, we preferred to take the long path. Actually, the techniques used for the abstract framework are useful for problems that do not fit in the convex reflexive setting, but where convexity and reflexivity still play an important role.

2.1 Extended Real-Valued Functions

In order to deal with unconstrained and constrained optimization problems in a unified setting, we introduce the *extended real numbers* by adding the *symbol* $+\infty$. The convention $\gamma < +\infty$ for all $\gamma \in \mathbf{R}$ allows us to extend the total order of $\mathbf{R}$ to a total order of $\mathbf{R} \cup \{+\infty\}$. We can define functions on a set X with *values* in $\mathbf{R} \cup \{+\infty\}$. The simplest example is the *indicator function* of a set $C \subset X$, defined as

$$\delta_C(x) = \begin{cases} 0 & \text{if } x \in C \\ +\infty & \text{otherwise.} \end{cases}$$

The main interest of introducing these kinds of functions is that, clearly, if $f : X \to \mathbf{R}$, the optimization problems

$$\min\{f(x) : x \in C\} \qquad \text{and} \qquad \min\{f(x) + \delta_C(x) : x \in X\}$$

J. Peypouquet, *Convex Optimization in Normed Spaces*,
SpringerBriefs in Optimization, DOI 10.1007/978-3-319-13710-0_2

are equivalent. The advantage of the second formulation is that linear, geometric, or topological properties of the underlying space X may be exploited.

Let $f : X \to \mathbf{R} \cup \{+\infty\}$ be an extended real-valued function. The *effective domain* (or, simply, the *domain*) of f is the set of points where f is finite. In other words,

$$\mathrm{dom}(f) = \{x \in X \ : \ f(x) < +\infty\}.$$

A function f is *proper* if $\mathrm{dom}(f) \neq \emptyset$. The alternative, namely $f \equiv +\infty$, is not very interesting for optimization purposes. Given $\gamma \in \mathbf{R}$, the *γ-sublevel set* of f is

$$\Gamma_\gamma(f) = \{x \in X : f(x) \leq \gamma\}.$$

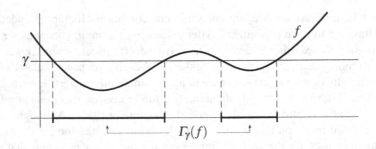

If $x \in \mathrm{dom}(f)$, then $x \in \Gamma_{f(x)}(f)$. Therefore, $\mathrm{dom}(f) = \bigcup_{\gamma \in \mathbf{R}} \Gamma_\gamma(f)$. On the other hand, recall that

$$\mathrm{argmin}(f) = \{x^* \in X \ : \ f(x^*) \leq f(x) \text{ for all } x \in X\}.$$

Observe that

$$\mathrm{argmin}(f) = \bigcap_{\gamma > \inf(f)} \Gamma_\gamma(f). \tag{2.1}$$

Finally, the *epigraph* of f is the subset of the product space $X \times \mathbf{R}$ defined as

$$\mathrm{epi}(f) = \{(x, \alpha) \in X \times \mathbf{R} \ : \ f(x) \leq \alpha\}.$$

This set includes the graph of f and all the points above it.

2.2 Lower-Semicontinuity and Minimization

In this section, we discuss the existence of minimizers for lower-semicontinuous functions in a fairly abstract setting.

Minimization in Hausdorff Spaces

In what follows, (X, τ) is a *Hausdorff space* (a topological space in which every two distinct points admit disjoint neighborhoods). A function $f : X \to \mathbf{R} \cup \{+\infty\}$ is *lower-semicontinuous* at a point $x_0 \in X$ if for each $\alpha < f(x_0)$ there is a neighborhood $\mathscr{V}$ of x_0 such that $f(y) > \alpha$ for all $y \in \mathscr{V}$. If f is lower-semicontinuous at every point of X, we say f is lower-semicontinuous in X.

Example 2.1. The indicator function δ_C of a closed set C is lower-semicontinuous.
□

Example 2.2. If f and g are lower-semicontinuous and if $\alpha \geq 0$, then $f + \alpha g$ is lower-semicontinuous. In other words, the set of lower-semicontinuous functions is a convex cone.
□

Lower-semicontinuity can be characterized in terms of the epigraph and the level sets of the function:

Proposition 2.3. *Let* $f : X \to \mathbf{R} \cup \{+\infty\}$. *The following are equivalent:*

i) The function f is lower-semicontinuous;
ii) The set $\mathrm{epi}(f)$ *is closed in* $X \times \mathbf{R}$; *and*
iii) For each $\gamma \in \mathbf{R}$, *the sublevel set* $\Gamma_\gamma(f)$ *is closed.*

Proof. We shall prove that i) $\Rightarrow$ ii) $\Rightarrow$ iii) $\Rightarrow$ i).

Let f be lower-semicontinuous and take $(x_0, \alpha) \notin \mathrm{epi}(f)$. We have $\alpha < f(x_0)$. Now take any $\beta \in (\alpha, f(x_0))$. There is a neighborhood $\mathscr{V}$ of x_0 such that $f(y) > \beta$ for all $y \in \mathscr{V}$. The set $\mathscr{V} \times (-\infty, \beta)$ is a neighborhood of (x_0, α) that does not intersect $\mathrm{epi}(f)$, which gives ii).

Suppose now that $\mathrm{epi}(f)$ is closed. For each $\gamma \in \mathbf{R}$, the set $\Gamma_\gamma(f)$ is homeomorphic to $\mathrm{epi}(f) \cap [X \times \{\gamma\}]$ and so it is closed.

Finally, assume $\Gamma_\gamma(f)$ is closed. In order to prove that f is lower-semicontinuous, pick $x_0 \in X$ and $\alpha \in \mathbf{R}$ such that $\alpha < f(x_0)$. Then $x_0 \notin \Gamma_\alpha(f)$. Since this set is closed, there is a neighborhood $\mathscr{V}$ of x_0 that does not intersect $\Gamma_\alpha(f)$. In other words, $f(y) > \alpha$ for all $y \in \mathscr{V}$.
□

Example 2.4. If $(f_i)_{i \in I}$ is a family of lower-semicontinuous functions, then $\sup(f_i)$ is lower-semicontinuous, since $\mathrm{epi}(\sup(f_i)) = \cap \mathrm{epi}(f_i)$ and the intersection of closed sets is closed.
□

A function $f : X \to \mathbf{R} \cup \{+\infty\}$ is *inf-compact* if, for each $\gamma \in \mathbf{R}$, the sublevel set $\Gamma_\gamma(f)$ is relatively compact. Clearly, a function $f : X \to \mathbf{R} \cup \{+\infty\}$ is proper, lower-semicontinuous, and inf-compact if, and only if, $\Gamma_\gamma(f)$ is nonempty and compact for each $\gamma > \inf(f)$.

Theorem 2.5. *Let* (X, τ) *be a Hausdorff space and let* $f : X \to \mathbf{R} \cup \{+\infty\}$ *be proper, lower-semicontinuous, and inf-compact. Then* $\mathrm{argmin}(f)$ *is nonempty and compact. Moreover,* $\inf(f) > -\infty$.

Proof. Let (γ_n) be a nonincreasing sequence such that $\lim\limits_{n\to\infty} \gamma_n = \inf(f)$ (possibly $-\infty$). For each $n \in \mathbf{N}$ consider the compact set $K_n = \Gamma_{\gamma_n}(f)$. Then $K_{n+1} \subset K_n$ for each $n \in \mathbf{N}$ and so

$$\operatorname{argmin}(f) = \bigcap_{n\in\mathbf{N}} K_n$$

is nonempty and compact. Clearly, $\inf(f) > -\infty$.   □

The usefulness of Theorem 2.5 relies strongly on the appropriate choice of topology for X. If τ has too many open sets, it will be very likely for a function $f : X \to \mathbf{R} \cup \{+\infty\}$ to be lower-semicontinuous but one would hardly expect it to be inf-compact. We shall see in, Sect. 2.3, that for convex functions defined on a reflexive space, the weak topology is a suitable choice.

Improving Approximate Minimizers

The following result provides an important geometric property satisfied by approximate minimizers of lower-semicontinuous functions:

Theorem 2.6 (Ekeland's Variational Principle). *Consider a lower-semicontinuous function $f : X \to \mathbf{R} \cup \{+\infty\}$ defined on a Banach space X. Let $\varepsilon > 0$ and suppose $x_0 \in \operatorname{dom}(f)$ is such that $f(x_0) \leq \inf\limits_{x\in X} f(x) + \varepsilon$. Then, for each $\lambda > 0$, there exists $\bar{x} \in \bar{B}(x_0, \varepsilon/\lambda)$ such that $f(\bar{x}) + \lambda\|\bar{x} - x_0\| \leq f(x_0)$ and $f(x) + \lambda\|x - \bar{x}\| > f(\bar{x})$ for all $x \neq \bar{x}$.*

Proof. For each $n \geq 0$, given $x_n \in X$, define

$$C_n = \{x \in X : f(x) + \lambda\|x - x_n\| \leq f(x_n)\}$$

and

$$v_n = \inf\{f(x) : x \in C_n\},$$

and take $x_{n+1} \in C_n$ such that

$$f(x_{n+1}) \leq v_n + \frac{1}{n+1}.$$

Observe that (C_n) is a nested (decreasing) sequence of closed sets, and $f(x_n)$ is nonincreasing and bounded from below by $f(x_0) - \varepsilon$. Clearly, for each $h \geq 0$ we have

$$\lambda\|x_{n+h} - x_n\| \leq f(x_n) - f(x_{n+h}),$$

and (x_n) is a Cauchy sequence. Its limit, which we denote $\bar{x}$, belongs to $\bigcap\limits_{n\geq 0} C_n$. In particular,

$$f(\bar{x}) + \lambda\|\bar{x} - x_0\| \leq f(x_0)$$

and

$$\lambda\|\bar{x} - x_0\| \leq f(x_0) - f(\bar{x}) \leq \varepsilon.$$

Finally, if there is $\tilde{x} \in X$ such that

$$f(\tilde{x}) + \lambda \|\tilde{x} - \bar{x}\| \le f(\bar{x}),$$

then, also $\tilde{x} \in \bigcap_{n \ge 0} C_n$ and so

$$f(\tilde{x}) + \lambda \|\tilde{x} - x_n\| \le f(x_n) \le f(\tilde{x}) + \frac{1}{n}$$

for all n, and we deduce that $\tilde{x} = \bar{x}$.   □

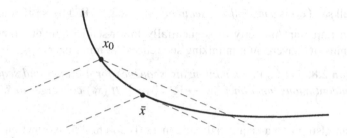

Minimizing Sequences

It is often useful to consider a sequential notion of lower-semicontinuity. A function $f : X \to \mathbf{R} \cup \{+\infty\}$ is *sequentially lower-semicontinuous* at $x \in \mathrm{dom}(f)$ for the topology τ if

$$f(x) \le \liminf_{n \to \infty} f(x_n)$$

for every sequence (x_n) converging to x for the topology τ.

Example 2.7. Let $f : \mathbf{R}^M \to \mathbf{R} \cup \{+\infty\}$ be proper, lower-semicontinuous, and bounded from below. Define $F : L^p(0,T;\mathbf{R}^M) \to \mathbf{R} \cup \{+\infty\}$ by

$$F(u) = \int_0^T f(u(t))\,dt.$$

We shall see that F is sequentially lower-semicontinuous. Let (u_n) be a sequence in the domain of F converging strongly to some $\bar{u} \in L^p(0,T;\mathbf{R}^M)$. Extract a subsequence (u_{k_n}) such that

$$\lim_{n \to \infty} F(u_{k_n}) = \liminf_{n \to \infty} F(u_n).$$

Next, since (u_{k_n}) converges in $L^p(0,T;\mathbf{R}^M)$, we may extract yet another subsequence (u_{j_n}) such that $(u_{j_n}(t))$ converges to $\bar{u}(t)$ for almost every $t \in [0,T]$ (see, for instance, [30, Theorem 4.9]). Since f is lower-semicontinuous, we must have

$\liminf\limits_{n\to\infty} f(u_{j_n}(t)) \geq f(\bar{u}(t))$ for almost every t. By Fatou's Lemma (see, for instance, [30, Lemma 4.1]), we obtain

$$F(\bar{u}) \leq \liminf_{n\to\infty} F(u_{j_n}) = \lim_{n\to\infty} F(u_{k_n}) = \liminf_{n\to\infty} F(u_n),$$

and so F is sequentially lower-semicontinuous. $\square$

It is possible to establish a sequential analogue of Proposition 2.3, along with relationships between lower-semicontinuity and sequential lower-semicontinuity for the strong and the weak topologies, in the spirit of Proposition 1.23. We shall come back to this point in Proposition 2.17.

We shall say (x_n) is a *minimizing sequence* for $f : X \to \mathbf{R} \cup \{+\infty\}$ if $\lim\limits_{n\to\infty} f(x_n) = \inf(f)$. An important property of sequentially lower-semicontinuous functions is that the limits of convergent minimizing sequences are minimizers.

Proposition 2.8. *Let (x_n) be a minimizing sequence for a proper and sequentially lower-semicontinuous function $f : X \to \mathbf{R} \cup \{+\infty\}$. If (x_n) converges to $\bar{x}$, then $\bar{x} \in \operatorname{argmin}(f)$.*

One can also prove a sequential version of Theorem 2.5. A function $f : X \to \mathbf{R} \cup \{+\infty\}$ is *sequentially inf-compact* if for each $\gamma > \inf(f)$, every sequence in $\Gamma_\gamma(f)$ has a convergent subsequence. We have the following:

Theorem 2.9. *If $f : X \to \mathbf{R} \cup \{+\infty\}$ is proper, sequentially lower-semicontinuous, and sequentially inf-compact, then there exists a convergent minimizing sequence for f. In particular, $\operatorname{argmin}(f)$ is nonempty.*

The proofs of Proposition 2.8 and Theorem 2.9 are straightforward, and left to the reader.

2.3 Minimizers of Convex Functions

An extended real valued function $f : X \to \mathbf{R} \cup \{+\infty\}$ defined on a vector space X is *convex* if

$$f(\lambda x + (1-\lambda)y) \leq \lambda f(x) + (1-\lambda)f(y) \tag{2.2}$$

for each $x, y \in \operatorname{dom}(f)$ and $\lambda \in (0,1)$. Notice that inequality (2.2) holds trivially if $\lambda \in \{0,1\}$ or if either x or y are not in $\operatorname{dom}(f)$. If the inequality in (2.2) is strict whenever $x \neq y$ and $\lambda \in (0,1)$, we say f is *strictly convex*. Moreover, f is *strongly convex* with parameter $\alpha > 0$ if

$$f(\lambda x + (1-\lambda)y) \leq \lambda f(x) + (1-\lambda)f(y) - \frac{\alpha}{2}\lambda(1-\lambda)\|x-y\|^2$$

for each $x, y \in \operatorname{dom}(f)$ and $\lambda \in (0,1)$.

Remark 2.10. It is easy to prove that f is convex if, and only if, epi(f) is a convex subset of $X \times \mathbf{R}$. Moreover, if f is convex, then each sublevel set $\Gamma_\gamma(f)$ is convex. Obviously the converse is not true in general: Simply consider any nonconvex monotone function on $X = \mathbf{R}$. Functions whose level sets are convex are called *quasi-convex*. $\qquad\square$

We shall provide some practical characterizations of convexity for differentiable functions in Sect. 3.2.

Example 2.11. Let $B : X \times X \to \mathbf{R}$ be a bilinear function and define $f(x) = B(x,x)$, as we did in Example 1.27. For each $x, y \in X$ and $\lambda \in (0,1)$, we have

$$f(\lambda x + (1-\lambda)y) = \lambda f(x) + (1-\lambda)f(y) - \lambda(1-\lambda)B(x-y, x-y).$$

Therefore, we have the following:

i) f is convex if, and only if, B is positive semidefinite ($B(z,z) \geq 0$ for all $z \in H$);
ii) f is strictly convex if, and only if, B is positive definite ($B(z,z) > 0$ for all $z \neq 0$); and
iii) f is strongly convex with parameter α if, and only if, B is uniformly elliptic with parameter α ($B(z,z) \geq \frac{\alpha}{2}\|z\|^2$).

In particular, if $A : H \to H$ is a linear operator on a Hilbert space H and we set $B(x,x) = \langle Ax, x \rangle$, then B is bilinear. Therefore, the function $f : H \to \mathbf{R}$, defined by $f(x) = \langle Ax, x \rangle$, is convex if, and only if, A is positive semidefinite ($\langle Az, z \rangle \geq 0$ for all $z \in H$); strictly convex if, and only if, A is positive definite ($\langle Az, z \rangle > 0$ for all $z \neq 0$); and strongly convex with parameter α if, and only if, A is uniformly elliptic with parameter α ($\langle Az, z \rangle \geq \frac{\alpha}{2}\|z\|^2$). $\qquad\square$

Example 2.12. The indicator function δ_C of a convex set C is a convex function. $\quad\square$

Some Convexity-Preserving Operations

We mention some operations that allow us to construct convex functions from others. Another very important example will be studied in detail in Sect. 3.6

Example 2.13. Suppose $A : X \to Y$ is affine, $f : Y \to \mathbf{R} \cup \{+\infty\}$ is convex and $\theta : \mathbf{R} \to \mathbf{R} \cup \{+\infty\}$ convex and nondecreasing. Then, the function $g = \theta \circ f \circ A : X \to \mathbf{R} \cup \{+\infty\}$ is convex. $\qquad\square$

Example 2.14. If f and g are convex and if $\alpha \geq 0$, then $f + \alpha g$ is convex. It follows that the set of convex functions is a convex cone. $\qquad\square$

Example 2.15. If $(f_i)_{i \in I}$ is a family of convex functions, then $\sup(f_i)$ is convex, since epi$(\sup(f_i)) = \cap$ epi(f_i) and the intersection of convex sets is convex. $\quad\square$

Example 2.16. In general, the infimum of convex functions need not be convex. However, we have the following: Let V be a vector space and let $g : X \times V \to \mathbf{R} \cup \{+\infty\}$ be convex. Then, the function $f : X \to \mathbf{R} \cup \{+\infty\}$ defined by $f(x) = \inf_{v \in V} g(x,v)$

is convex. Here, the facts that g is convex in the product space and the infimum is taken over a whole vector space are crucial.   □

Convexity, Lower-Semicontinuity, and Existence of Minimizers

As a consequence of Propositions 1.21, 1.23 and 2.3, we obtain

Proposition 2.17. *Let $(X, \| \cdot \|)$ be a normed space and let $f : X \to \mathbf{R} \cup \{+\infty\}$. Consider the following statements:*

i) f is weakly lower-semicontinuous.
ii) f is weakly sequentially lower-semicontinuous.
iii) f is sequentially lower-semicontinuous.
iv) f is lower-semicontinuous.

Then i) $\Rightarrow$ ii) $\Rightarrow$ iii) $\Leftrightarrow$ iv) $\Leftarrow$ i). If f is convex, the four statements are equivalent.

Example 2.18. Let $f : \mathbf{R}^M \to \mathbf{R} \cup \{+\infty\}$ be proper, convex, lower-semicontinuous, and bounded from below. As in Example 2.7, define $F : L^p(0, T; \mathbf{R}^M) \to \mathbf{R} \cup \{+\infty\}$ by

$$F(u) = \int_0^T f(u(t)) \, dt.$$

Clearly, F is proper and convex. Moreover, we already proved that F is sequentially lower-semicontinuous. Proposition 2.17 shows that F is lower-semicontinuous and sequentially lower-semicontinuous both for the strong and the weak topologies.   □

A function $f : X \to \mathbf{R} \cup \{+\infty\}$ is *coercive* if

$$\lim_{\|x\| \to \infty} f(x) = \infty,$$

or, equivalently, if $\Gamma_\gamma(f)$ is bounded for each $\gamma \in \mathbf{R}$. By Theorem 1.24, coercive functions on reflexive spaces are weakly inf-compact. We have the following:

Theorem 2.19. *Let X be reflexive. If $f : X \to \mathbf{R} \cup \{+\infty\}$ is proper, convex, coercive, and lower-semicontinuous, then $\operatorname{argmin}(f)$ is nonempty and weakly compact. If, moreover, f is strictly convex, then $\operatorname{argmin}(f)$ is a singleton.*

Proof. The function f fulfills the hypotheses of Theorem 2.5 for the weak topology. Clearly, a strictly convex function cannot have more than one minimizer.   □

If f is strongly convex, it is strictly convex and coercive. Therefore, we deduce:

Corollary 2.20. *Let X be reflexive. If $f : X \to \mathbf{R} \cup \{+\infty\}$ is proper, strongly convex, and lower-semicontinuous, then $\operatorname{argmin}(f)$ is a singleton.*

Chapter 3
Convex Analysis and Subdifferential Calculus

Abstract This chapter deals with several properties of convex functions, especially in connection with their regularity, on the one hand, and the characterization of their minimizers, on the other. We shall explore sufficient conditions for a convex function to be continuous, as well as several connections between convexity and differentiability. Next, we present the notion of subgradient, a generalization of the concept of derivative for nondifferentiable convex functions that will allow us to characterize their minimizers. After discussing conditions that guarantee their existence, we present the basic (yet subtle) calculus rules, along with their remarkable consequences. Other important theoretical and practical tools, such as the Fenchel conjugate and the Lagrange multipliers, will also be studied. These are particularly useful for solving constrained problems.

3.1 Convexity and Continuity

In this section, we discuss some characterizations of continuity and lower-semi-continuity for convex functions.

Lower-Semicontinuous Convex Functions

Pretty much as closed convex sets are intersections of closed half-spaces, any lower-semicontinuous convex function can be represented as a supremum of continuous affine functions:

Proposition 3.1. *Let $f : X \to \mathbf{R} \cup \{+\infty\}$ be proper. Then, f is convex and lower-semicontinuous if, and only if, there exists a family $(f_i)_{i \in I}$ of continuous affine functions on X such that $f = \sup(f_i)$.*

J. Peypouquet, *Convex Optimization in Normed Spaces*,
SpringerBriefs in Optimization, DOI 10.1007/978-3-319-13710-0_3

Proof. Suppose f is convex and lower-semicontinuous and let $x_0 \in X$. We shall prove that, for every $\lambda_0 < f(x_0)$, there exists a continuous affine function α such that $\alpha(x) \leq f(x)$ for all $x \in \text{dom}(f)$ and $\lambda_0 < \alpha(x_0) < f(x_0)$. Since $\text{epi}(f)$ is nonempty, closed, and convex, and $(x_0, \lambda_0) \notin \text{epi}(f)$, by the Hahn–Banach Separation Theorem 1.10, there exist $(L, s) \in X^* \times \mathbf{R} \setminus \{(0,0)\}$ and $\varepsilon > 0$ such that

$$\langle L, x_0 \rangle + s\lambda_0 + \varepsilon \leq \langle L, x \rangle + s\lambda \tag{3.1}$$

for all $(x, \lambda) \in \text{epi}(f)$. Clearly, $s \geq 0$. Otherwise, we may take $x \in \text{dom}(f)$ and λ sufficiently large to contradict (3.1). We distinguish two cases:

$\underline{s > 0}$: We may assume, without loss of generality, that $s = 1$ (or divide by s and rename L and ε). We set

$$\alpha(x) = \langle -L, x \rangle + [\langle L, x_0 \rangle + \lambda_0 + \varepsilon],$$

and take $\lambda = f(x)$ to deduce that $f(x) \geq \alpha(x)$ for all $x \in \text{dom}(f)$ and $\alpha(x_0) > \lambda_0$. Observe that this is valid for each $x_0 \in \text{dom}(f)$.

$\underline{s = 0}$: As said above, necessarily $x_0 \notin \text{dom}(f)$, and so, $f(x_0) = +\infty$. Set

$$\alpha_0(x) = \langle -L, x \rangle + [\langle L, x_0 \rangle + \varepsilon],$$

and observe that $\alpha_0(x) \leq 0$ for all $x \in \text{dom}(f)$ and $\alpha_0(x_0) = \varepsilon > 0$. Now take $\hat{x} \in \text{dom}(f)$ and use the argument of the case $s > 0$ to obtain a continuous affine function $\hat{\alpha}$ such that $f(x) \geq \hat{\alpha}(x)$ for all $x \in \text{dom}(f)$. Given $n \in \mathbf{N}$ set $\alpha_n = \hat{\alpha} + n\alpha_0$. We conclude that $f(x) \geq \alpha_n(x)$ for all $x \in \text{dom}(f)$ and $\lim_{n \to \infty} \alpha_n(x_0) = \lim_{n \to \infty} (\hat{\alpha}(x_0) + n\varepsilon) = +\infty = f(x_0)$.

The converse is straightforward, since $\text{epi}(\sup(f_i)) = \cap \text{epi}(f_i)$. $\square$

Characterization of Continuity

In this subsection, we present some continuity results that reveal how remarkably regular convex functions are.

Proposition 3.2. *Let $f : X \to \mathbf{R} \cup \{+\infty\}$ be convex and fix $x_0 \in X$. The following are equivalent:*

i) f is bounded from above in a neighborhood of x_0;
ii) f is Lipschitz-continuous in a neighborhood of x_0;
iii) f is continuous in x_0; and
iv) $(x_0, \lambda) \in \text{int}(\text{epi}(f))$ for each $\lambda > f(x_0)$.

Proof. We shall prove that i) $\Rightarrow$ ii) $\Rightarrow$ iii) $\Rightarrow$ iv) $\Rightarrow$ i).

i) $\Rightarrow$ ii): There exist $r > 0$ and $K > f(x_0)$ such that $f(z) \leq K$ for every $z \in B(x_0, 2r)$. We shall find $M > 0$ such that $|f(x) - f(y)| \leq M\|x - y\|$ for all $x, y \in$

$B(x_0, r)$. First, consider the point

$$\tilde{y} = y + r\frac{y - x}{\|y - x\|}. \tag{3.2}$$

Since $\|\tilde{y} - x_0\| \leq \|y - x_0\| + r < 2r$, we have $f(\tilde{y}) \leq K$. Solving for y in (3.2), we see that

$$y = \lambda\tilde{y} + (1 - \lambda)x, \quad \text{with} \quad \lambda = \frac{\|y - x\|}{\|y - x\| + r} \leq \frac{\|y - x\|}{r}.$$

The convexity of f implies

$$f(y) - f(x) \leq \lambda[f(\tilde{y}) - f(x)] \leq \lambda[K - f(x)]. \tag{3.3}$$

Write $x_0 = \frac{1}{2}x + \frac{1}{2}(2x_0 - x)$. We have $f(x_0) \leq \frac{1}{2}f(x) + \frac{1}{2}f(2x_0 - x)$ and

$$-f(x) \leq K - 2f(x_0). \tag{3.4}$$

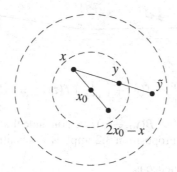

Combining (3.3) and (3.4), we deduce that

$$f(y) - f(x) \leq 2\lambda[K - f(x_0)] \leq \frac{2(K - f(x_0))}{r}\|x - y\|.$$

Interchanging the roles of x and y we conclude that $|f(x) - f(y)| \leq M\|x - y\|$ with $M = \frac{2(K - f(x_0))}{r} > 0$.

ii) $\Rightarrow$ iii): This is straightforward.

iii) $\Rightarrow$ iv): If f is continuous in x_0 and $\lambda > f(x_0)$, then for each $\eta \in (f(x_0), \lambda)$ there exists $\delta > 0$ such that $f(z) < \eta$ for all $z \in B(x_0, \delta)$. Hence, $(x_0, \lambda) \in B(x_0, \delta) \times (\eta, +\infty) \subset \text{epi}(f)$, and $(x_0, \lambda) \in \text{int}(\text{epi}(f))$.

iv) $\Rightarrow$ i): If $(x_0, \lambda) \in \text{int}(\text{epi}(f))$, then there exist $r > 0$ and $K > f(x_0)$, such that $B(x_0, r) \times (K, +\infty) \subset \text{epi}(f)$. It follows that $f(z) \leq K$ for every $z \in B(x_0, r)$. $\square$

Proposition 3.3. *Let $(X, \|\cdot\|)$ be a normed space and let $f : X \to \mathbf{R} \cup \{+\infty\}$ be convex. If f is continuous at some $x_0 \in \text{dom}(f)$, then $x_0 \in \text{int}(\text{dom}(f))$ and f is continuous on $\text{int}(\text{dom}(f))$.*

Proof. If f is continuous at x_0, Proposition 3.2 gives $x_0 \in \text{int}(\text{dom}(f))$ and there exist $r > 0$ y $K > f(x_0)$ such that $f(x) \leq K$ for every $x \in B(x_0, r)$. Let $y_0 \in$

int(dom(f)) and pick $\rho > 0$ such that the point $z_0 = y_0 + \rho(y_0 - x_0)$ belongs to dom(f). Take $y \in B(y_0, \frac{\rho r}{1+\rho})$ and set $w = x_0 + (\frac{1+\rho}{\rho})(y - y_0)$. Solving for y, we see that

$$y = \left(\frac{\rho}{1+\rho}\right) w + \left(\frac{1}{1+\rho}\right) z_0.$$

On the other hand, $\|w - x_0\| = \left(\frac{1+\rho}{\rho}\right)\|y - y_0\| < r$, and so $w \in B(x_0, r)$ y $f(w) \le K$.

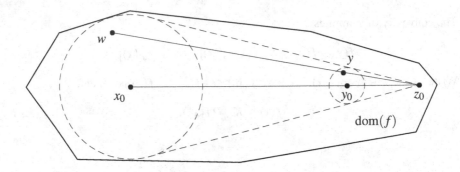

Therefore,

$$f(y) \le \left(\tfrac{\rho}{1+\rho}\right) f(w) + \left(\tfrac{1}{1+\rho}\right) f(z_0) \le \max\{K, f(z_0)\}.$$

Since this is true for each $y \in B(y_0, \frac{\rho r}{1+\rho})$, we conclude that f is bounded from above in a neighborhood of y_0. Proposition 3.2 implies f is continuous at y_0. □

An immediate consequence is:

Corollary 3.4. *Let* $f : X \to \mathbf{R} \cup \{+\infty\}$ *be convex. If* int($\Gamma_\lambda(f)$) $\ne \emptyset$ *for some* $\lambda \in \mathbf{R}$, *then* f *is continuous on* int(dom(f)).

Proposition 3.3 requires that f be continuous at some point. This hypothesis can be waived if the space is complete. We begin by establishing the result in finite-dimensional spaces:

Proposition 3.5. *Let* X *be finite dimensional and let* $f : X \to \mathbf{R} \cup \{+\infty\}$ *be convex. Then* f *is continuous on* int(dom(f)).

Proof. Let $\{e_1, \ldots, e_N\}$ generate X. Let $x_0 \in$ int(dom(f)) and take $\rho > 0$ small enough so that $x_0 \pm \rho e_i \in$ dom(f) for all $i = 1, \ldots, N$. The convex hull C of these points is a neighborhood of x_0 and f is bounded by $\max_{i=1,\ldots,N}\{f(x_0 \pm \rho e_i)\}$ on C. The result follows from Proposition 3.2. □

For general Banach spaces, we have the following:

Proposition 3.6. *Let* $(X, \|\cdot\|)$ *be a Banach space and let* $f : X \to \mathbf{R} \cup \{+\infty\}$ *be lower-semicontinuous and convex. Then* f *is continuous on* int(dom(f)).

Proof. Fix $x_0 \in \text{int}(\text{dom}(f))$. Without any loss of generality we may assume that $x_0 = 0$. Take $\lambda > f(0)$. Given $x \in X$, define $g_x : \mathbf{R} \to \mathbf{R} \cup \{+\infty\}$ by $g_x(t) = f(tx)$. Since $0 \in \text{int}(\text{dom}(g_x))$, we deduce that g_x is continuous at 0 by Proposition 3.5. Therefore, there is $t_x > 0$ such that $t_x x \in \Gamma_\lambda(f)$. Repeating this argument for each $x \in X$, we see that $\bigcup_{n \geq 1} n \Gamma_\lambda(f) = X$. Baire's Category Theorem shows that $\Gamma_\lambda(f)$ has nonempty interior and we conclude by Corollary 3.4.   □

Remark 3.7. To summarize, let $(X, \|\cdot\|)$ be a normed space and let $f : X \to \mathbf{R} \cup \{+\infty\}$ be convex. Then f is continuous on $\text{int}(\text{dom}(f))$ if either (i) f is continuous at some point, (ii) X is finite dimensional, or (iii) X is a Banach space and f is lower-semicontinuous.   □

3.2 Convexity and Differentiability

In this subsection, we study some connections between convexity and differentiability. First, we analyze the existence and properties of directional derivatives. Next, we provide characterizations for the convexity of differentiable functions. Finally, we provide equivalent conditions for the gradient of a differentiable convex function to be Lipschitz continuous.

3.2.1 Directional Derivatives

Recall from Sect. 1.1.4 that the directional derivative of a function $f : X \to \mathbf{R} \cup \{+\infty\}$ at a point $x \in \text{dom}(f)$ in the direction $h \in X$ is given by $f'(x;h) = \lim\limits_{t \to 0} \frac{f(x+th)-f(x)}{t}$.

Remark 3.8. If f is convex, a simple computation shows that the quotient $\frac{f(x+th)-f(x)}{t}$ is nondecreasing as a function of t. We deduce that

$$f'(x;h) = \inf_{t>0} \frac{f(x+th)-f(x)}{t},$$

which *exists* in $\mathbf{R} \cup \{-\infty, +\infty\}$.   □

We have the following:

Proposition 3.9. *Let* $f : X \to \mathbf{R} \cup \{+\infty\}$ *be proper and convex, and let* $x \in \text{dom}(f)$. *Then, the function* $\phi_x : X \to [-\infty, +\infty]$, *defined by* $\phi_x(h) = f'(x;h)$, *is convex. If, moreover,* f *is continuous in* x, *then* ϕ_x *is finite and continuous in* X.

Proof. Take $y, z \in X$, $\lambda \in (0,1)$ and $t > 0$. Write $h = \lambda y + (1-\lambda)z$. By the convexity of f, we have

$$\frac{f(x+th)-f(x)}{t} \leq \lambda \frac{f(x+ty)-f(x)}{t} + (1-\lambda)\frac{f(x+tz)-f(x)}{t}.$$

Passing to the limit, $\phi_x(\lambda y+(1-\lambda)z) \leq \lambda\phi_x(y)+(1-\lambda)\phi_x(z)$. Now, if f is continuous in x, it is Lipschitz-continuous in a neighborhood of x, by Proposition 3.2. Then, for all $h \in X$, all sufficiently small $t > 0$ and some $L > 0$, we have

$$-L\|h\| \leq \frac{f(x+th)-f(x)}{t} \leq L\|h\|.$$

It follows that $\mathrm{dom}(\phi_x) = X$ and that ϕ_x is bounded from above in a neighborhood of 0. Using Proposition 3.2 again, we deduce that ϕ_x is continuous in 0, and, by Proposition 3.3, ϕ_x is continuous in $\mathrm{int}(\mathrm{dom}(\phi_x)) = X$.   □

A function f is Gâteaux-differentiable at x if the function ϕ_x defined above is *linear* and *continuous* in X. Proposition 3.9 provides the continuity part, but it is clear that a continuous convex function need not be Gâteaux-differentiable (take, for instance, the absolute value in $\mathbf{R}$). A sufficient condition for a convex function to be Gâteaux-differentiable will be given in Proposition 3.58.

3.2.2 Characterizations of Convexity for Differentiable Functions

We begin by providing the following characterization for the convexity of Gâteaux-differentiable functions:

Proposition 3.10 (Characterization of convexity). *Let $A \subset X$ be open and convex, and let $f : A \to \mathbf{R}$ be Gâteaux-differentiable. The following are equivalent:*

i) f is convex.
ii) $f(y) \geq f(x)+\langle \nabla f(x),y-x\rangle$ for every $x,y \in A$.
iii) $\langle \nabla f(x)-\nabla f(y),x-y\rangle \geq 0$ for every $x,y \in A$.

If, moreover, f is twice Gâteaux-differentiable on A, then the preceding conditions are equivalent to

iv) $\langle \nabla^2 f(x)d,d\rangle \geq 0$ for every $x \in A$ and $d \in X$ (positive semidefinite).

Proof. By convexity,

$$f(\lambda y+(1-\lambda)x) \leq \lambda f(y)+(1-\lambda)f(x)$$

for all $y \in X$ and all $\lambda \in (0,1)$. Rearranging the terms we get

$$\frac{f(x+\lambda(y-x))-f(x)}{\lambda} \leq f(y)-f(x).$$

As $\lambda \to 0$ we obtain *ii)*. From *ii)*, we immediately deduce *iii)*.
To prove that *iii)* $\Rightarrow$ *i)*, define $\phi : [0,1] \to \mathbf{R}$ by

$$\phi(\lambda) = f(\lambda x+(1-\lambda)y)-\lambda f(x)-(1-\lambda)f(y).$$

Then $\phi(0) = \phi(1) = 0$ and

$$\phi'(\lambda) = \langle \nabla f\left(\lambda x + (1-\lambda)y\right), x-y\rangle - f(x) + f(y)$$

for $\lambda \in (0,1)$. Take $0 < \lambda_1 < \lambda_2 < 1$ and write $x_i = \lambda_i x + (1-\lambda_i)y$ for $i = 1,2$. A simple computation shows that

$$\phi'(\lambda_1) - \phi'(\lambda_2) = \frac{1}{\lambda_1 - \lambda_2}\langle \nabla f(x_1) - \nabla f(x_2), x_1 - x_2\rangle \le 0.$$

In other words, ϕ' is nondecreasing. Since $\phi(0) = \phi(1) = 0$, there is $\bar{\lambda} \in (0,1)$ such that $\phi'(\bar{\lambda}) = 0$. Since ϕ' is nondecreasing, $\phi' \le 0$ (and ϕ is nonincreasing) on $[0, \bar{\lambda}]$ and next $\phi' \ge 0$ (whence ϕ is nondecreasing) on $[\bar{\lambda}, 1]$. It follows that $\phi(\lambda) \le 0$ on $[0,1]$, and so, f is convex.

Assume now that f is twice Gâteaux-differentiable and let us prove that $iii) \Rightarrow iv) \Rightarrow i)$.

For $t > 0$ and $h \in X$, we have $\langle \nabla f(x+th) - \nabla f(x), th\rangle \ge 0$. Dividing by t^2 and passing to the limit as $t \to 0$, we obtain $\langle \nabla^2 f(x)h, h\rangle \ge 0$.

Finally, defining ϕ as above, we see that

$$\phi''(\lambda) = \langle \nabla^2 f\left(\lambda x + (1-\lambda)y\right)(x-y), x-y\rangle \ge 0.$$

It follows that ϕ' is nondecreasing and we conclude as before. $\square$

It is possible to obtain characterizations for strict and strong convexity as well. The details are left as an exercise.

Proposition 3.11 (Characterization of strict convexity). *Let $A \subset X$ be open and convex, and let $f : A \to \mathbf{R}$ be Gâteaux-differentiable. The following are equivalent:*

i) f is strictly convex.
ii) $f(y) > f(x) + \langle \nabla f(x), y-x\rangle$ for any distinct $x, y \in A$.
iii)$\langle \nabla f(x) - \nabla f(y), x-y\rangle > 0$ for any distinct $x, y \in A$.

If, moreover, f is twice Gâteaux-differentiable on A, then the following condition is sufficient for the previous three:

iv) $\langle \nabla^2 f(x)d, d\rangle > 0$ for every $x \in A$ and $d \in X$ (positive definite).

Proposition 3.12 (Characterization of strong convexity). *Let $A \subset X$ be open and convex, and let $f : A \to \mathbf{R}$ be Gâteaux-differentiable. The following are equivalent:*

i) f is strongly convex with constant $\alpha > 0$.
ii) $f(y) \ge f(x) + \langle \nabla f(x), y-x\rangle + \frac{\alpha}{2}\|x-y\|^2$ for every $x, y \in A$.
iii)$\langle \nabla f(x) - \nabla f(y), x-y\rangle \ge \alpha\|x-y\|^2$ for every $x, y \in A$.

If, moreover, f is twice Gâteaux-differentiable on A, then the preceding conditions are equivalent to

iv) $\langle \nabla^2 f(x)d, d\rangle \ge \frac{\alpha}{2}\|d\|^2$ for every $x \in A$ and $d \in X$ (uniformly elliptic with constant α).

3.2.3 Lipschitz-Continuity of the Gradient

A function $F : X \to X^*$ is *cocoercive* with constant β if

$$\langle F(x) - F(y), x - y \rangle \geq \beta \|F(x) - F(y)\|_*^2$$

for all $x, y \in X$.

Theorem 3.13 (Baillon–Haddad Theorem). *Let $f : X \to \mathbf{R}$ be convex and differentiable. The gradient of f is Lipschitz-continuous with constant L if, and only if, it is cocoercive with constant $1/L$.*

Proof. Let $x, y, z \in X$. Suppose ∇f is Lipschitz-continuous with constant L. By the Descent Lemma 1.30, we have

$$f(z) \leq f(x) + \langle \nabla f(x), z - x \rangle + \frac{L}{2}\|z - x\|^2.$$

Subtract $\langle \nabla f(y), z \rangle$ to both sides, write $h_y(z) = f(z) - \langle \nabla f(y), z \rangle$ and rearrange the terms to obtain

$$h_y(z) \leq h_x(x) + \langle \nabla f(x) - \nabla f(y), z \rangle + \frac{L}{2}\|z - x\|^2.$$

The function h_y is convex, differentiable and $\nabla h_y(y) = 0$. We deduce that $h_y(y) \leq h_y(z)$ for all $z \in X$, and so

$$h_y(y) \leq h_x(x) + \langle \nabla f(x) - \nabla f(y), z \rangle + \frac{L}{2}\|z - x\|^2. \tag{3.5}$$

Fix any $\varepsilon > 0$ and pick $\mu \in X$ such that $\|\mu\| \leq 1$ and

$$\langle \nabla f(x) - \nabla f(y), \mu \rangle \geq \|\nabla f(x) - \nabla f(y)\|_* - \varepsilon.$$

Set $R = \frac{\|\nabla f(x) - \nabla f(y)\|}{L}$ and replace $z = x - R\mu$ in the right-hand side of (3.5) to obtain

$$h_y(y) \leq h_x(x) + \langle \nabla f(x) - \nabla f(y), x \rangle - \frac{\|\nabla f(x) - \nabla f(y)\|^2}{2L} + R\varepsilon.$$

Interchanging the roles of x and y and adding the resulting inequality, we obtain

$$\frac{1}{L}\|\nabla f(x) - \nabla f(y)\|^2 \leq \langle \nabla f(x) - \nabla f(y), x - y \rangle + 2R\varepsilon.$$

Since this holds for each $\varepsilon > 0$, we deduce that ∇f is cocoercive with constant $1/L$. The converse is straightforward. $\qquad\square$

Remark 3.14. From the proof of the Baillon–Haddad Theorem 3.13, we see that the L-Lipschitz continuity and the $\frac{1}{L}$ cocoercivity of ∇f are actually equivalent to the validity of the inequality given by the Descent Lemma 1.30, whenever f is convex.

3.3 Subgradients, Subdifferential and Fermat's Rule

Let $f : X \to \mathbf{R}$ be convex and assume it is Gâteaux-differentiable at a point $x \in X$. By convexity, Proposition 3.10 gives

$$f(y) \geq f(x) + \langle \nabla f(x), y - x \rangle \tag{3.6}$$

for each $y \in X$. This shows that the hyperplane

$$V = \{(y,z) \in X \times \mathbf{R} : f(x) + \langle \nabla f(x), y - x \rangle = z\}$$

lies below the set $\mathrm{epi}(f)$ and touches it at the point $(x, f(x))$.

This idea can be generalized to nondifferentiable functions. Let $f : X \to \mathbf{R} \cup \{+\infty\}$ be proper and convex. A point $x^* \in X^*$ is a *subgradient* of f at x if

$$f(y) \geq f(x) + \langle x^*, y - x \rangle \tag{3.7}$$

for all $y \in X$, or, equivalently, if (3.7) holds for all y in a neighborhood of x. The set of all subgradients of f at x is the *subdifferential* of f at x and is denoted by $\partial f(x)$. If $\partial f(x) \neq \emptyset$, we say f is *subdifferentiable* at x. The *domain* of the subdifferential is

$$\mathrm{dom}(\partial f) = \{x \in X : \partial f(x) \neq \emptyset\}.$$

By definition, $\mathrm{dom}(\partial f) \subset \mathrm{dom}(f)$. The inclusion may be strict though, as in Example 3.17 below. However, we shall see (Corollary 3.34) $\overline{\mathrm{dom}(\partial f)} = \overline{\mathrm{dom}(f)}$.

Let us see some examples:

Example 3.15. For $f : \mathbf{R} \to \mathbf{R}$, given by $f(x) = |x|$, we have $\partial f(x) = \{-1\}$ if $x < 0$, $\partial f(0) = [-1, 1]$, and $\partial f(x) = \{1\}$ for $x > 0$. □

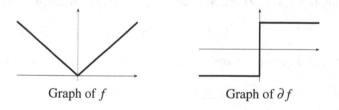

Graph of f Graph of ∂f

Example 3.16. More generally, if $f : X \to \mathbf{R}$ is given by $f(x) = \|x - x_0\|$, with $x_0 \in X$, then

$$\partial f(x) = \begin{cases} B_{X^*}(0, 1) & \text{if } x = x_0 \\ \mathscr{F}(x - x_0) & \text{if } x \neq x_0, \end{cases}$$

where $\mathscr{F}$ is the normalized duality mapping defined in Sect. 1.1.2. □

Example 3.17. Define $g : \mathbf{R} \to \mathbf{R} \cup \{+\infty\}$ by $g(x) = +\infty$ if $x < 0$, and $g(x) = -\sqrt{x}$ if $x \geq 0$. Then

$$\partial g(x) = \begin{cases} \emptyset & \text{if } x \leq 0 \\ \left\{-\frac{1}{2\sqrt{x}}\right\} & \text{if } x > 0. \end{cases}$$

Notice that $0 \in \mathrm{dom}(g)$ but $\partial g(0) = \emptyset$, thus $\mathrm{dom}(\partial g) \subsetneq \mathrm{dom}(g)$. □

Example 3.18. Let $h : \mathbf{R} \to \mathbf{R} \cup \{+\infty\}$ be given by $h(x) = +\infty$ if $x \neq 0$, and $h(0) = 0$. Then

$$\partial h(x) = \begin{cases} \emptyset & \text{if } x \neq 0 \\ \mathbf{R} & \text{if } x = 0. \end{cases}$$

Observe that h is subdifferentiable but not continuous at 0. □

Example 3.19. More generally, let C be a nonempty, closed, and convex subset of X and let $\delta_C : X \to \mathbf{R} \cup \{+\infty\}$ be the indicator function of C:

$$\delta_C(x) = \begin{cases} 0 & \text{if } x \in C \\ +\infty & \text{otherwise.} \end{cases}$$

Given $x \in X$, the set $N_C(x) = \partial \delta_C(x)$ is the *normal cone* to C at x. It is given by

$$N_C(x) = \{x^* \in X^* : \langle x^*, y - x \rangle \leq 0 \text{ for all } y \in C\}$$

if $x \in C$, and $N_C(x) = \emptyset$ if $x \notin C$. Intuitively, the normal cone contains the directions that point *outwards* with respect to C. If C is a closed affine subspace, that is $C = \{x_0\} + V$, where $x_0 \in X$ and V is a closed subspace of X. In this case, $N_C(x) = V^\perp$ for all $x \in C$ (here $V^\perp$ is the orthogonal space of V, see Sect. 1.1.1). □

The subdifferential really is an extension of the notion of derivative:

Proposition 3.20. *Let $f : X \to \mathbf{R} \cup \{+\infty\}$ be convex. If f is Gâteaux-differentiable at x, then $x \in \mathrm{dom}(\partial f)$ and $\partial f(x) = \{\nabla f(x)\}$.*

Proof. First, the gradient inequality (3.6) and the definition of the subdifferential together imply $\nabla f(x) \in \partial f(x)$. Now take any $x^* \in \partial f(x)$. By definition,

$$f(y) \geq f(x) + \langle x^*, y - x \rangle$$

for all $y \in X$. Take any $h \in X$ and $t > 0$, and write $y = x + th$ to deduce that

$$\frac{f(x + th) - f(x)}{t} \geq \langle x^*, h \rangle.$$

Passing to the limit as $t \to 0$, we deduce that $\langle \nabla f(x) - x^*, h \rangle \geq 0$. Since this must be true for each $h \in X$, necessarily $x^* = \nabla f(x)$. □

Proposition 3.58 provides a converse result.

A convex function may have more than one subgradient. However, we have:

Proposition 3.21. *For each $x \in X$, the set $\partial f(x)$ is closed and convex.*

Proof. Let $x_1^*, x_2^* \in \partial f(x)$ and $\lambda \in (0,1)$. For each $y \in X$ we have

$$f(y) \geq f(x) + \langle x_1^*, y - x \rangle$$
$$f(y) \geq f(x) + \langle x_2^*, y - x \rangle$$

Add λ times the first inequality and $1 - \lambda$ times the second one to obtain

$$f(y) \geq f(x) + \langle \lambda x_1^* + (1 - \lambda) x_2^*, y - x \rangle.$$

Since this holds for each $y \in X$, we have $\lambda x_1^* + (1 - \lambda) x_2^* \in \partial f(x)$. To see that $\partial f(x)$ is (sequentially) closed, take a sequence (x_n^*) in $\partial f(x)$, converging to some x^*. We have

$$f(y) \geq f(x) + \langle x_n^*, y - x \rangle$$

for each $y \in X$ and $n \in \mathbf{N}$. As $n \to \infty$ we obtain

$$f(y) \geq f(x) + \langle x^*, y - x \rangle.$$

It follows that $x^* \in \partial f(x)$. □

Recall that if $f : \mathbf{R} \to \mathbf{R}$ is convex and differentiable, then its derivative is non-decreasing. The following result generalizes this fact:

Proposition 3.22. *Let $f : X \to \mathbf{R} \cup \{+\infty\}$ be convex. If $x^* \in \partial f(x)$ and $y^* \in \partial f(y)$, then $\langle x^* - y^*, x - y \rangle \geq 0$.*

Proof. Note that $f(y) \geq f(x) + \langle x^*, y - x \rangle$ and $f(x) \geq f(y) + \langle y^*, x - y \rangle$. It suffices to add these inequalities and rearrange the terms. □

This property is known as *monotonicity*.

For strongly convex functions, we have the following result:

Proposition 3.23. *Let $f : X \to \mathbf{R} \cup \{+\infty\}$ be strongly convex with parameter α. Then, for each $x^* \in \partial f(x)$ and $y \in X$, we have*

$$f(y) \geq f(x) + \langle x^*, y - x \rangle + \frac{\alpha}{2} \|x - y\|^2.$$

Moreover, for each $y^* \in \partial f(y)$, we have $\langle x^* - y^*, x - y \rangle \geq \alpha \|x - y\|^2$.

The definition of the subdifferential has a straightforward yet remarkable consequence, namely:

Theorem 3.24 (Fermat's Rule). *Let $f : X \to \mathbf{R} \cup \{+\infty\}$ be proper and convex. Then $\hat{x}$ is a global minimizer of f if, and only if, $0 \in \partial f(\hat{x})$.*

For convex functions, the condition $0 \in \partial f(\hat{x})$ given by Fermat's Rule is not only necessary but also sufficient for $\hat{x}$ to be a global minimizer. In particular, in view of

Proposition 3.20, if f is convex and differentiable, then $0 = \nabla f(\hat{x})$ if, and only if, $\hat{x}$ is a global minimizer of f. In other words, the only critical points convex functions may have are their global minimizers.

3.4 Subdifferentiablility

As pointed out in Example 3.18, subdifferentiability does not imply continuity. Surprisingly enough, the converse is true.

Proposition 3.25. *Let $f : X \to \mathbf{R} \cup \{+\infty\}$ be convex. If f is continuous at x, then $\partial f(x)$ is nonempty and bounded.*

Proof. According to Proposition 3.2, $\text{int}(\text{epi}(f)) \neq \emptyset$. Part i) of the Hahn–Banach Separation Theorem 1.10, with $A = \text{int}(\text{epi}(f))$ and $B = \{(x, f(x))\}$, gives $(L, s) \in X^* \times \mathbf{R} \setminus \{(0,0)\}$ such that

$$\langle L, y \rangle + s\lambda \leq \langle L, x \rangle + s f(x)$$

for every $(y, \lambda) \in \text{int}(\text{epi}(f))$. Taking $y = x$, we deduce that $s \leq 0$. If $s = 0$ then $\langle L, y - x \rangle \leq 0$ for every y in a neighborhood of x. Hence $L = 0$, which is a contradiction. We conclude that $s < 0$. Therefore,

$$\lambda \geq f(x) + \langle z^*, y - x \rangle,$$

with $z^* = -L/s$, for every $\lambda > f(y)$. Letting λ tend to $f(y)$, we see that

$$f(y) \geq f(x) + \langle z^*, y - x \rangle.$$

This implies $z^* \in \partial f(x) \neq \emptyset$. On the other hand, since f is continuous at x, it is Lipschitz-continuous on a neighborhood of x, by Proposition 3.2. If $x^* \in \partial f(x)$, then

$$f(x) + \langle x^*, y - x \rangle \leq f(y) \leq f(x) + M\|y - x\|,$$

and so, $\langle x^*, y - x \rangle \leq M\|y - x\|$ for every y in a neighborhood of x. We conclude that $\|x^*\| \leq M$. In other words, $\partial f(x) \subset \bar{B}(0, M)$. $\qquad\square$

Approximate Subdifferentiability

As seen in Example 3.17, it may happen that $\text{dom}(\partial f) \subsetneq \text{dom}(f)$. This can be an inconvenient when trying to implement optimization algorithms. Given $\varepsilon > 0$, a point $x^* \in X^*$ is an *ε-approximate subgradient* of f at $x \in \text{dom}(f)$ if

$$f(y) \geq f(x) + \langle x^*, y - x \rangle - \varepsilon \tag{3.8}$$

for all $y \in X$ (compare with the subdifferential inequality (3.7)). The set $\partial_\varepsilon f(x)$ of such vectors is the *ε-approximate subdifferential* of f at x. Clearly, $\partial f(x) \subset \partial_\varepsilon f(x)$ for each x, and so

$$\text{dom}(\partial f) \subset \text{dom}(\partial_\varepsilon f) \subset \text{dom}(f)$$

for each $\varepsilon > 0$. It turns out that the last inclusion is, in fact, an equality.

Proposition 3.26. *Let $f : X \to \mathbf{R} \cup \{+\infty\}$ be proper, convex, and lower-semicontinuous. Then $\text{dom}(f) = \text{dom}(\partial_\varepsilon f)$ for all $\varepsilon > 0$.*

Proof. Let $\varepsilon > 0$ and let $x \in \text{dom}(f)$. By Proposition 3.1, f can be represented as the pointwise supremum of continuous affine functions. Therefore, there exist $x^* \in X^*$ and $\mu \in \mathbf{R}$, such that

$$f(y) \geq \langle x^*, y \rangle + \mu$$

for all $y \in X$ and

$$\langle x^*, x \rangle + \mu \geq f(x) - \varepsilon.$$

Adding these two inequalities, we obtain precisely (3.8). □

Example 3.27. Let us recall from Example 3.17 that the function $g : \mathbf{R} \to \mathbf{R} \cup \{+\infty\}$ given by $g(x) = +\infty$ if $x < 0$, and $g(x) = -\sqrt{x}$ if $x \geq 0$; satisfies $0 \in \text{dom}(g)$ but $\partial g(0) = \emptyset$. A simple computation shows that $\partial_\varepsilon g(0) = \left(-\infty, \frac{1}{4\varepsilon}\right]$ for $\varepsilon > 0$. Observe that $\text{dom}(\partial g) \subsetneq \text{dom}(\partial_\varepsilon g) = \text{dom}(g)$. □

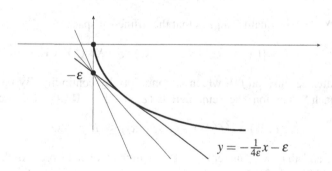

$$y = -\frac{1}{4\varepsilon}x - \varepsilon$$

3.5 Basic Subdifferential Calculus Rules and Applications

In this section, we discuss some calculus rules for the subdifferentials of convex functions. We begin by providing a *Chain Rule* for the composition with a bounded linear operator. Next, we present the Moreau–Rockafellar Theorem, regarding the subdifferential of a sum. This result has several important consequences, both for theoretical and practical purposes.

3.5.1 Composition with a Linear Function: A Chain Rule

The following is a consequence of the Hahn–Banach Separation Theorem 1.10:

Proposition 3.28 (Chain Rule). *Let $A \in \mathscr{L}(X;Y)$ and let $f : Y \to \mathbf{R} \cup \{+\infty\}$ be proper, convex, and lower-semicontinuous. For each $x \in X$, we have*

$$A^* \partial f(Ax) \subset \partial(f \circ A)(x).$$

Equality holds for every $x \in X$ if f is continuous at some $y_0 \in A(X)$.

Proof. Let $x \in X$ and $x^* \in \partial f(Ax)$. For every $y \in Y$, we have

$$f(y) \geq f(Ax) + \langle x^*, y - Ax \rangle.$$

In particular,

$$f(Az) \geq f(Ax) + \langle x^*, A(z - x) \rangle$$

for every $z \in X$. We conclude that

$$(f \circ A)(z) \geq (f \circ A)(x) + \langle A^* x^*, z - x \rangle$$

for each $z \in X$, and so $A^* \partial f(Ax) \subseteq \partial(f \circ A)(x)$.

Conversely, take $x^* \in \partial(f \circ A)(x)$, which means that

$$f(Az) \geq f(Ax) + \langle x^*, z - x \rangle$$

for all $z \in X$. This inequality implies that the affine subspace

$$V = \{(Az, f(Ax) + \langle x^*, z - x \rangle) : z \in X\} \subset Y \times \mathbf{R}$$

does not intersect $\operatorname{int}(\operatorname{epi}(f))$, which is nonempty by continuity. By part i) of the Hahn–Banach Separation Theorem, there is $(L, s) \in Y^* \times \mathbf{R} \setminus \{(0,0)\}$ such that

$$\langle L, Az \rangle + s(f(Ax) + \langle x^*, z - x \rangle) \leq \langle L, y \rangle + s\lambda$$

for all $z \in X$ and all $(y, \lambda) \in \operatorname{int}(\operatorname{epi}(f))$. Pretty much like in the proof of the Moreau–Rockafellar Theorem 3.30, we prove that $s > 0$. Then, write $\ell = -L/s$ and let $\lambda \to f(y)$ to obtain

$$f(y) \geq f(Ax) + \langle x^*, z - x \rangle + \langle \ell, y - Az \rangle. \tag{3.9}$$

Take $z = x$ to deduce that

$$f(y) \geq f(Ax) + \langle \ell, y - Ax \rangle.$$

for all $y \in Y$. On the other hand, if we take $y = Ax$ in (3.9), we obtain

$$0 \geq \langle x^* - A^* \ell, z - x \rangle$$

for al $z \in X$. We conclude that $\ell \in \partial f(Ax)$ and $x^* = A^* \ell$, whence $x^* \in A^* \partial f(Ax)$. □

3.5.2 Sum of Convex Functions and the Moreau–Rockafellar Theorem

A natural question is whether the subdifferential of the sum of two functions, is the sum of their subdifferentials. We begin by showing that this is not always the case.

Example 3.29. Let $f,g: \mathbf{R} \to \mathbf{R} \cup \{+\infty\}$ be given by

$$f(x) = \begin{cases} 0 & \text{if } x \le 0 \\ +\infty & \text{if } x > 0 \end{cases} \quad \text{and} \quad g(x) = \begin{cases} +\infty & \text{if } x < 0 \\ -\sqrt{x} & \text{if } x \ge 0. \end{cases}$$

We have

$$\partial f(x) = \begin{cases} \{0\} & \text{if } x < 0 \\ [0,+\infty) & \text{if } x = 0 \\ \emptyset & \text{if } x > 0 \end{cases} \quad \text{and} \quad \partial g(x) = \begin{cases} \emptyset & \text{if } x \le 0 \\ \left\{-\frac{1}{2\sqrt{x}}\right\} & \text{if } x > 0. \end{cases}$$

Therefore, $\partial f(x) + \partial g(x) = \emptyset$ for every $x \in \mathbf{R}$. On the other hand, $f + g = \delta_{\{0\}}$, which implies $\partial(f+g)(x) = \emptyset$ if $x \ne 0$, but $\partial(f+g)(0) = \mathbf{R}$. We see that $\partial(f+g)(x)$ may differ from $\partial f(x) + \partial g(x)$. □

Roughly speaking, the problem with the function in the example above is that the intersection of their domains is too small. We have the following:

Theorem 3.30 (Moreau–Rockafellar Theorem). *Let $f,g: X \to \mathbf{R} \cup \{+\infty\}$ be proper, convex, and lower-semicontinuous. For each $x \in X$, we have*

$$\partial f(x) + \partial g(x) \subset \partial(f+g)(x). \tag{3.10}$$

Equality holds for every $x \in X$ if f is continuous at some $x_0 \in \text{dom}(g)$.

Proof. If $x^* \in \partial f(x)$ and $z^* \in \partial g(x)$, then

$$f(y) \ge f(x) + \langle x^*, y-x \rangle \quad \text{and} \quad g(y) \ge g(x) + \langle z^*, y-x \rangle$$

for each $y \in X$. Adding both inequalities, we conclude that

$$f(y) + g(y) \ge f(x) + g(x) + \langle x^* + z^*, y-x \rangle$$

for each $y \in X$ and so, $x^* + z^* \in \partial(f+g)(x)$.

Suppose now that $u^* \in \partial(f+g)(x)$. We have

$$f(y) + g(y) \ge f(x) + g(x) + \langle u^*, y-x \rangle \tag{3.11}$$

for every $y \in X$. We shall find $x^* \in \partial f(x)$ and $z^* \in \partial g(x)$ such that $x^* + z^* = u^*$. To this end, consider the following nonempty convex sets:

$$B = \{(y,\lambda) \in X \times \mathbf{R} : g(y) - g(x) \le -\lambda\}$$
$$C = \{(y,\lambda) \in X \times \mathbf{R} : f(y) - f(x) - \langle u^*, y-x \rangle \le \lambda\}.$$

Define $h : X \to \mathbf{R} \cup \{+\infty\}$ as $h(y) = f(y) - f(x) - \langle u^*, y - x \rangle$. Since h is continuous in x_0 and $C = \mathrm{epi}(h)$, the open convex set $A = \mathrm{int}(C)$ is nonempty by Proposition 3.2. Moreover, $A \cap B = \emptyset$ by inequality (3.11). Using Hahn–Banach Separation Theorem 1.10, we obtain $(L, s) \in X^* \times \mathbf{R} \setminus \{(0, 0)\}$ such that

$$\langle L, y \rangle + s\lambda \leq \langle L, z \rangle + s\mu$$

for each $(y, \lambda) \in A$ and each $(z, \mu) \in B$. In particular, taking $(y, \lambda) = (x, 1) \in A$ and $(z, \mu) = (x, 0) \in B$, we deduce that $s \leq 0$. On the other hand, if $s = 0$, taking $z = x_0$ we see that $\langle L, x_0 - y \rangle \geq 0$ for every y in a neighborhood of x_0. This implies $L = 0$ and contradicts the fact that $(L, s) \neq (0, 0)$. Therefore, $s < 0$, and we may write

$$\langle z^*, y \rangle + \lambda \leq \langle z^*, z \rangle + \mu \tag{3.12}$$

with $z^* = -L/s$. By the definition of C, taking again $(z, \mu) = (x, 0) \in B$, we obtain

$$\langle z^*, y - x \rangle + f(y) - f(x) - \langle u^*, y - x \rangle \leq 0.$$

Inequality (3.11) then gives

$$g(y) \geq g(x) + \langle z^*, y - x \rangle$$

for every $y \in \mathrm{dom}(g)$, and we conclude that $z^* \in \partial g(x)$. In a similar fashion, use (3.12) with $y = x$, any $\lambda > 0$ and $\mu = g(x) - g(z)$, along with (3.11), to deduce that

$$f(z) \geq f(x) + \langle u^* - z^*, z - x \rangle$$

for all $z \in X$, which implies $x^* = u^* - z^* \in \partial f(x)$ and completes the proof. $\qquad\square$

Observe that, if $\partial(f + g)(x) = \partial f(x) + \partial g(x)$ for all $x \in X$, then $\mathrm{dom}(\partial(f + g)) = \mathrm{dom}(\partial f) \cap \mathrm{dom}(\partial g)$.

If X is a Banach space, another condition ensuring the equality in (3.10) is that $\bigcup_{t \geq 0} t \left(\mathrm{dom}(f) - \mathrm{dom}(g) \right)$ be a closed subset of X. This is known as the *Attouch–Brézis condition* [9].

3.5.3 Some Consequences

Combining the Chain Rule (Proposition 3.28) and the Moreau–Rockafellar Theorem 3.30, we obtain the following:

Corollary 3.31. *Let $A \in \mathscr{L}(X; Y)$, and let $f : Y \to \mathbf{R} \cup \{+\infty\}$ and $f, g : X \to \mathbf{R} \cup \{+\infty\}$ be proper, convex, and lower-semicontinuous. For each $x \in X$, we have*

$$A^* \partial f(Ax) + \partial g(x) \subset \partial(f \circ A + g)(x).$$

Equality holds for every $x \in X$ if there is $x_0 \in \mathrm{dom}(g)$ such that f is continuous at Ax_0.

This result will be useful in the context of Fenchel–Rockafellar duality (see Theorem 3.51).

Another interesting consequence of the Moreau–Rockafellar Theorem 3.30 is:

Corollary 3.32. *Let X be reflexive and let $f : X \to \mathbf{R} \cup \{+\infty\}$ be proper, strongly convex, and lower-semicontinuous. Then $\partial f(X) = X^*$. If, moreover, f is Gâteaux-differentiable, then $\nabla f : X \to X^*$ is a bijection.*

Proof. Take $x^* \in X^*$. The function $g : X \to \mathbf{R} \cup \{+\infty\}$, defined by $g(x) = f(x) - \langle x^*, x \rangle$, is proper, strongly convex, and lower-semicontinuous. By Theorem 2.19, there is a unique $\bar{x} \in \mathrm{argmin}(g)$ and, by Fermat's Rule (Theorem 3.24), $0 \in \partial g(\bar{x})$. This means that $x^* \in \partial f(\bar{x})$. If f is Gâteaux-differentiable, the equation $\nabla f(x) = x^*$ has exactly one solution for each $x^* \in X^*$. $\qquad \Box$

Points in the graph of the approximate subdifferential can be approximated by points that are actually in the graph of the subdifferential.

Theorem 3.33 (Brønsted–Rockafellar Theorem). *Let X be a Banach space and let $f : X \to \mathbf{R} \cup \{+\infty\}$ be lower-semicontinuous and convex. Take $\varepsilon > 0$ and $x_0 \in \mathrm{dom}(f)$, and assume $x_0^* \in \partial_\varepsilon f(x_0)$. Then, for each $\lambda > 0$, there exist $\bar{x} \in \mathrm{dom}(\partial f)$ and $\bar{x}^* \in \partial f(\bar{x})$ such that $\|\bar{x} - x_0\| \leq \varepsilon / \lambda$ and $\|\bar{x}^* - x_0^*\|_* \leq \lambda$.* $\qquad \Box$

Proof. Since $x_0^* \in \partial_\varepsilon f(x_0)$, we have

$$f(y) \geq f(x_0) + \langle x^*, y - x_0 \rangle - \varepsilon$$

for all $y \in X$. In other words, x_0 satisfies

$$g(x_0) \leq \inf_{y \in X} g(y) + \varepsilon,$$

where we have written $g(y) = f(y) - \langle x_0^*, y \rangle$. By Ekeland's Variational Principle (Theorem 2.6), there exists $\bar{x} \in B(x_0, \varepsilon / \lambda)$ such that

$$g(\bar{x}) + \lambda \|\bar{x} - x_0\| \leq g(x_0) \qquad \text{and} \qquad g(\bar{x}) < g(y) + \lambda \|\bar{x} - y\|$$

for all $y \neq \bar{x}$. The latter implies that $\bar{x}$ is the unique minimizer of the function $h : X \to \mathbf{R} \cup \{+\infty\}$ defined by $h(y) = g(y) + \lambda \|\bar{x} - y\|$. From the Moreau–Rockafellar Theorem 3.30, it follows that $\bar{x} \in \mathrm{dom}(\partial h) = \mathrm{dom}(\partial f)$ and

$$0 \in \partial h(\bar{x}) = \partial g(\bar{x}) + \lambda B = \partial f(\bar{x}) - x_0^* + \lambda B,$$

where $B = \{x^* \in X^* : \|x^*\|_* \leq 1\}$ (see also Example 3.16). In other words, there exists $\bar{x}^* \in \partial f(\bar{x})$ such that $\|\bar{x}^* - x_0^*\|_* \leq \lambda$.

An immediate consequence is the following:

Corollary 3.34. *Let X be a Banach space and let $f : X \to \mathbf{R} \cup \{+\infty\}$ be convex and lower-semicontinuous. Then $\overline{\mathrm{dom}(\partial f)} = \overline{\mathrm{dom}(f)}$.*

Proof. Since $\mathrm{dom}(\partial f) \subset \mathrm{dom}(f)$, it suffices to prove that $\mathrm{dom}(f) \subset \overline{\mathrm{dom}(\partial f)}$. Indeed, take $x_0 \in \mathrm{dom}(f)$ and let $\varepsilon > 0$. By the Brønsted–Rockafellar Theorem 3.33 (take $\lambda = 1$), there is $\bar{x} \in \mathrm{dom}(\partial f)$ such that $\|\bar{x} - x_0\| \leq \varepsilon$. □

In other words, the points of subdifferentiability are dense in the domain.

3.5.4 Moreau–Yosida Regularization and Smoothing

In this subsection, we present a regularizing and smoothing technique for convex functions defined on a Hilbert space. By adding a quadratic term, one is able to force the existence and uniqueness of a minimizer. This fact, in turn, has two major consequences: first, it is the core of an important minimization method, known as the proximal point algorithm, which we shall study in Chap. 6; and second, it allows us to construct a smooth version of the function having the same minimizers.

The Moreau–Yosida Regularization

Let H be a Hilbert space and let $f : H \to \mathbf{R} \cup \{+\infty\}$ be proper, convex and lower-semicontinuous. Given $\lambda > 0$ and $x \in H$, the *Moreau–Yosida Regularization* of f with parameter (λ, x) is the function $f_{(\lambda,x)} : H \to \mathbf{R} \cup \{+\infty\}$, defined by

$$f_{(\lambda,x)}(z) = f(z) + \frac{1}{2\lambda}\|z - x\|^2.$$

The following property will be useful later on, especially in Sect. 6.2, when we study the *proximal point algorithm* (actually, it is the core of that method):

Proposition 3.35. *For each $\lambda > 0$ and $x \in H$, the function $f_{(\lambda,x)}$ has a unique minimizer $\bar{x}$. Moreover, $\bar{x}$ is characterized by the inclusion*

$$-\frac{\bar{x} - x}{\lambda} \in \partial f(\bar{x}). \tag{3.13}$$

Proof. Since f is proper, convex, and lower-semicontinuous, $f_{(\lambda,x)}$ is proper, strictly convex, lower-semicontinuous, and coercive. Existence and uniqueness of a minimizer $\bar{x}$ is given by Theorem 2.19. Finally, Fermat's Rule (Theorem 3.24) and the Moreau–Rockafellar Theorem 3.30 imply that $\bar{x}$ must satisfy

$$0 \in \partial f_{(\lambda,x)}(\bar{x}) = \partial f(\bar{x}) + \frac{\bar{x} - x}{\lambda},$$

which gives the result. □

Solving for $\bar{x}$ in (3.13), we can write

$$\bar{x} = (I + \lambda \partial f)^{-1} x, \tag{3.14}$$

where $I : H \to H$ denotes the identity function. In view of Proposition 3.35, the expression $J_\lambda = (I + \lambda \partial f)^{-1}$ defines a function $J_\lambda : H \to H$, called the *proximity operator* of f, with parameter λ. It is also known as the *resolvent* of the operator ∂f with parameter λ, due to its analogy with the resolvent of a linear operator. We have the following:

Proposition 3.36. *If* $f : H \to \mathbf{R} \cup \{+\infty\}$ *is proper, lower-semicontinuous, and convex, then* $J_\lambda : H \to H$ *is (everywhere defined and) nonexpansive.*

Proof. Let $\bar{x} = J_\lambda(x)$ and $\bar{y} = J_\lambda(y)$, so that

$$-\frac{\bar{x} - x}{\lambda} \in \partial f(\bar{x}) \qquad \text{y} \qquad -\frac{\bar{y} - y}{\lambda} \in \partial f(\bar{y}).$$

In view of the monotonicity of ∂f (Proposition 3.22), we have

$$\langle (\bar{x} - x) - (\bar{y} - y), \bar{x} - \bar{y} \rangle \le 0.$$

Therefore,

$$0 \le \|\bar{x} - \bar{y}\|^2 \le \langle x - y, \bar{x} - \bar{y} \rangle \le \|x - y\| \, \|\bar{x} - \bar{y}\|, \tag{3.15}$$

and we conclude that $\|\bar{x} - \bar{y}\| \le \|x - y\|$. $\qquad\square$

Example 3.37. The indicator function δ_C of a nonempty, closed, and convex subset of H is proper, convex, and lower-semicontinuous. Given $\lambda > 0$ and $x \in H$, $J_\lambda(x)$ is the unique solution of

$$\min\left\{ \delta_C(z) + \frac{1}{2\lambda} \|z - x\|^2 : z \in H \right\} = \min\{ \|z - x\| : z \in C \}.$$

Independently of λ, $J_\lambda(x)$ is the point in C which is closest to x. In other words, it is the projection of x onto C, which we have denoted by $P_C(x)$. From Proposition 3.36, we recover the known fact that the function $P_C : H \to H$, defined by $P_C(x) = J_\lambda(x)$, is nonexpansive (see Proposition 1.40). $\qquad\square$

The Moreau Envelope and Its Smoothing Effect

Recall from Proposition 3.35 that for each $\lambda > 0$ and $x \in H$, the Moreau–Yosida Regularization $f_{(\lambda,x)}$ of f has a unique minimizer $\bar{x}$. The *Moreau envelope* of f with parameter $\lambda > 0$ is the function $f_\lambda : H \to \mathbf{R}$ defined by

$$f_\lambda(x) = \min_{z \in H}\{ f_{(\lambda,x)}(z) \} = \min_{z \in H}\left\{ f(z) + \frac{1}{2\lambda} \|z - x\|^2 \right\} = f(\bar{x}) + \frac{1}{2\lambda} \|\bar{x} - x\|^2.$$

An immediate consequence of the definition is that

$$\inf_{z \in H} f(z) \le f_\lambda(x) \le f(x)$$

for all $\lambda > 0$ and $x \in H$. Therefore,

$$\inf_{z \in H} f(z) = \inf_{z \in H} f_\lambda(z).$$

Moreover, $\hat{x}$ minimizes f on H if, and only if, it minimizes f_λ on H for all $\lambda > 0$.

Example 3.38. Let C be a nonempty, closed, and convex subset of H, and let $f = \delta_C$. Then, for each $\lambda > 0$ and $x \in H$, we have

$$f_\lambda(x) = \min_{z \in C} \left\{ \frac{1}{2\lambda} \|z - x\|^2 \right\} = \frac{1}{2\lambda} \text{dist}(x, C)^2,$$

where $\text{dist}(x, C)$ denotes the distance from x to C.   □

A remarkable property of the Moreau envelope is given by the following:

Proposition 3.39. *For each $\lambda > 0$, the function f_λ is Fréchet-differentiable and*

$$Df_\lambda(x) = \frac{1}{\lambda}(x - \bar{x})$$

for all $x \in H$. Moreover, f_λ is convex and Df_λ is Lipschitz-continuous with constant $1/\lambda$.

Proof. As before, write $\bar{x} = J_\lambda(x)$, $\bar{y} = J_\lambda(y)$, and observe that

$$f_\lambda(y) - f_\lambda(x) = f(\bar{y}) - f(\bar{x}) + \frac{1}{2\lambda} \left[\|\bar{y} - y\|^2 - \|\bar{x} - x\|^2 \right]$$

$$\ge -\frac{1}{\lambda} \langle \bar{x} - x, \bar{y} - \bar{x} \rangle + \frac{1}{2\lambda} \left[\|\bar{y} - y\|^2 - \|\bar{x} - x\|^2 \right],$$

by the subdifferential inequality and the inclusion (3.13). Straightforward algebraic manipulations yield

$$f_\lambda(y) - f_\lambda(x) - \frac{1}{\lambda} \langle x - \bar{x}, y - x \rangle \ge \frac{1}{2\lambda} \|(\bar{y} - \bar{x}) - (y - x)\|^2 \ge 0. \tag{3.16}$$

Interchanging the roles of x and y, we obtain

$$f_\lambda(x) - f_\lambda(y) - \frac{1}{\lambda} \langle y - \bar{y}, x - y \rangle \ge 0.$$

We deduce that

$$0 \le f_\lambda(y) - f_\lambda(x) - \frac{1}{\lambda} \langle x - \bar{x}, y - x \rangle \le \frac{1}{\lambda} \left[\|y - x\|^2 - \langle \bar{y} - \bar{x}, y - x \rangle \right] \le \frac{1}{\lambda} \|y - x\|^2,$$

in view of (3.15). Writing $y = x + h$, we conclude that

$$\lim_{\|h\| \to 0} \frac{1}{\|h\|} \left| f_\lambda(x+h) - f_\lambda(x) - \frac{1}{\lambda} \langle x - \bar{x}, h \rangle \right| = 0,$$

and f_λ is Fréchet-differentiable. The convexity follows from the fact that

$$\langle Df_\lambda(x) - Df_\lambda(y), x - y \rangle \geq 0,$$

and the characterization given in Proposition 3.10. Finally, from (3.15), we deduce that

$$\|(x - \bar{x}) - (y - \bar{y})\|^2 = \|x - y\|^2 + \|\bar{x} - \bar{y}\|^2 - 2\langle x - y, \bar{x} - \bar{y} \rangle \leq \|x - y\|^2.$$

Therefore,

$$\|Df_\lambda(x) - Df_\lambda(y)\| \leq \frac{1}{\lambda} \|x - y\|,$$

and so, Df_λ is Lipschitz-continuous with constant $1/\lambda$. ☐

Remark 3.40. Observe that $Df_\lambda(x) \in \partial f(J_\lambda(x))$. ☐

The approximation property of the Moreau envelope is given by the following:

Proposition 3.41. *As $\lambda \to 0$, f_λ converges pointwise to f.*

Proof. Since

$$f(J_\lambda(x)) + \frac{1}{2\lambda} \|J_\lambda(x) - x\|^2 \leq f(x)$$

and f is bounded from below by some continuous affine function (Proposition 3.1), $J_\lambda(x)$ remains bounded as $\lambda \to 0$. As a consequence, $f(J_\lambda(x))$ is bounded from below, and so, $\lim_{\lambda \to 0} \|J_\lambda(x) - x\| = 0$. Finally,

$$f(x) \leq \liminf_{\lambda \to 0} f(J_\lambda(x)) \leq \limsup_{\lambda \to 0} f(J_\lambda(x)) \leq f(x),$$

by lower-semicontinuity. ☐

3.6 The Fenchel Conjugate

The *Fenchel conjugate* of a proper function $f : X \to \mathbf{R} \cup \{+\infty\}$ is the function $f^* : X^* \to \mathbf{R} \cup \{+\infty\}$ defined by

$$f^*(x^*) = \sup_{x \in X} \{\langle x^*, x \rangle - f(x)\}. \tag{3.17}$$

Since f^* is a supremum of continuous affine functions, it is convex and lower-semicontinuous (see Proposition 3.1). Moreover, if f is bounded from below by a continuous affine function (for instance if f is proper and lower-semicontinuous), then f^* is proper.

3.6.1 Main Properties and Examples

As a consequence of the definition, we deduce the following:

Proposition 3.42 (Fenchel–Young Inequality). *Let $f : X \to \mathbf{R} \cup \{+\infty\}$. For all $x \in X$ and $x^* \in X^*$, we have*

$$f(x) + f^*(x^*) \geq \langle x^*, x \rangle. \tag{3.18}$$

If f is convex, then equality in (3.18) holds if, and only if, $x^* \in \partial f(x)$.

In particular, if f is convex, then $R(\partial f) \subset \mathrm{dom}(f^*)$. But the relationship between these two sets goes much further:

Proposition 3.43. *Let X be a Banach space and let $f : X \to \mathbf{R} \cup \{+\infty\}$ be lower-semicontinuous and convex, then $\overline{R(\partial f)} = \mathrm{dom}(f^*)$.*

Proof. Clearly, $R(\partial f) \subset \mathrm{dom}(f^*)$, so it suffices to prove that $\mathrm{dom}(f^*) \subset \overline{R(\partial f)}$. Let $x_0^* \in \mathrm{dom}(f^*)$ and let $\varepsilon > 0$. Take $x_0 \in \mathrm{dom}(f)$ such that

$$\langle x_0^*, x_0 \rangle - f(x_0) + \varepsilon \geq f^*(x_0^*) = \sup_{y \in X} \{ \langle x_0^*, y \rangle - f(y) \}.$$

We deduce that $x_0^* \in \partial_\varepsilon f(x_0)$. By the Brønsted–Rockafellar Theorem 3.33 (take $\lambda = \varepsilon$), there exist $\bar{x} \in \mathrm{dom}(\partial f)$ and $\bar{x}^* \in \partial f(\bar{x})$ such that $\|\bar{x} - x_0\| \leq 1$ and $\|\bar{x}^* - x_0^*\|_* \leq \varepsilon$. We conclude that $x_0^* \in \overline{R(\partial f)}$. $\square$

Another direct consequence of the Fenchel–Young Inequality (3.42) is:

Corollary 3.44. *If $f : X \to \mathbf{R}$ is convex and differentiable, then $\nabla f(X) \subset \mathrm{dom}(f^*)$ and $f^*(\nabla f(x)) = \langle \nabla f(x), x \rangle - f(x)$ for all $x \in X$.*

Let us see some examples:

Example 3.45. Let $f : \mathbf{R} \to \mathbf{R}$ be defined by $f(x) = e^x$. Then

$$f^*(x^*) = \begin{cases} x^* \ln(x^*) - x^* & \text{if } x^* > 0 \\ 0 & \text{if } x^* = 0 \\ +\infty & \text{if } x^* < 0. \end{cases}$$

This function is called the *Bolzmann–Shannon Entropy*. $\square$

Example 3.46. Let $f : X \to \mathbf{R}$ be defined by $f(x) = \langle x_0^*, x \rangle + \alpha$ for some $x_0^* \in X^*$ and $\alpha \in \mathbf{R}$. Then,

$$f^*(x^*) = \sup_{x \in X} \{ \langle x^* - x_0^*, x \rangle - \alpha \} = \begin{cases} +\infty & \text{if } x^* \neq x_0^* \\ -\alpha & \text{if } x^* = x_0^*. \end{cases}$$

In other words, $f^* = \delta_{\{x_0^*\}} - \alpha$.   □

Example 3.47. Let δ_C be the indicator function of a nonempty, closed convex subset C of X. Then,

$$\delta_C^*(x^*) = \sup_{x \in C} \{ \langle x^*, x \rangle \}.$$

This is the *support function* of the set C, and is denoted by $\sigma_C(x^*)$.   □

Example 3.48. The Fenchel conjugate of a radial function is radial: Let $\phi : [0, \infty) \to \mathbf{R} \cup \{+\infty\}$ and define $f : X \to \mathbf{R} \cup \{+\infty\}$ by $f(x) = \phi(\|x\|)$. Extend ϕ to all of $\mathbf{R}$ by $\phi(t) = +\infty$ for $t < 0$. Then

$$f^*(x^*) = \sup_{x \in X} \{ \langle x^*, x \rangle - \phi(\|x\|) \}$$

$$= \sup_{t \geq 0} \sup_{\|h\| = 1} \{ t \langle x^*, h \rangle - \phi(t) \}$$

$$= \sup_{t \in \mathbf{R}} \{ t \|x^*\|_* - \phi(t) \}$$

$$= \phi^*(\|x^*\|_*).$$

For instance, for $\phi(t) = \frac{1}{2} t^2$, we obtain $f^*(x^*) = \frac{1}{2} \|x^*\|_*^2$.   □

Example 3.49. Let $f : X \to \mathbf{R} \cup \{+\infty\}$ be proper and convex, and let $x \in \mathrm{dom}(f)$. For $h \in X$, set

$$\phi_x(h) = f'(x; h) = \lim_{t \to 0} \frac{f(x + th) - f(x)}{t} = \inf_{t > 0} \frac{f(x + th) - f(x)}{t}$$

(see Remark 3.8). Let us compute ϕ_x^*:

$$\phi_x^*(x^*) = \sup_{h \in X} \{ \langle x^*, h \rangle - \phi_x(h) \}$$

$$= \sup_{h \in X} \sup_{t > 0} \left\{ \langle x^*, h \rangle - \frac{f(x + th) - f(x)}{t} \right\}$$

$$= \sup_{t > 0} \sup_{z \in X} \left\{ \frac{f(x) + \langle x^*, z \rangle - f(z) - \langle x^*, x \rangle}{t} \right\}$$

$$= \sup_{t > 0} \left\{ \frac{f(x) + f^*(x^*) - \langle x^*, x \rangle}{t} \right\}.$$

By the Fenchel–Young Inequality (Proposition 3.42), we conclude that $\phi_x^*(x^*) = 0$ if $x^* \in \partial f(x)$, and $\phi_x^*(x^*) = +\infty$ otherwise. In other words, $\phi_x^*(x^*) = \delta_{\partial f(x)}(x^*)$.   □

Another easy consequence of the definition is:

Proposition 3.50. *If $f \leq g$, then $f^* \geq g^*$. In particular,*

$$\left(\sup_{i \in I} (f_i) \right)^* \leq \inf_{i \in I} (f_i^*) \qquad and \qquad \left(\inf_{i \in I} (f_i) \right)^* = \sup_{i \in I} (f_i^*)$$

for any family $(f_i)_{i \in I}$ of functions on X with values in $\mathbf{R} \cup \{+\infty\}$.

3.6.2 Fenchel–Rockafellar Duality

Let X and Y be normed spaces and let $A \in \mathscr{L}(X;Y)$. Consider two proper, lower-semicontinuous, and convex functions $f : X \to \mathbf{R} \cup \{+\infty\}$ and $g : Y \to \mathbf{R} \cup \{+\infty\}$. The *primal problem* in Fenchel–Rockafellar duality is given by:

$$(P) \qquad\qquad \inf_{x \in X} f(x) + g(Ax).$$

Its optimal value is denoted by α and the set of *primal solutions* is S. Also Consider the *dual problem*:

$$(D) \qquad\qquad \inf_{y^* \in Y^*} f^*(-A^*y^*) + g^*(y^*),$$

with optimal value α^*. The set of *dual solutions* is denoted by S^*.

By the Fenchel–Young Inequality (Proposition 3.42), for each $x \in X$ and $y^* \in Y^*$, we have $f(x) + f^*(-A^*y^*) \geq \langle -A^*y^*, x \rangle$ and $g(Ax) + g^*(y^*) \geq \langle y^*, Ax \rangle$. Thus,

$$f(x) + g(Ax) + f^*(-A^*y^*) + g^*(y^*) \geq \langle -A^*y^*, x \rangle + \langle y^*, Ax \rangle = 0 \qquad (3.19)$$

for all $x \in X$ and $y^* \in Y^*$, and so $\alpha + \alpha^* \geq 0$. The *duality gap* is $\alpha + \alpha^*$.

Let us characterize the primal-dual solutions:

Theorem 3.51. *The following statements concerning points $\hat{x} \in X$ and $\hat{y}^* \in Y^*$ are equivalent:*

i) $-A^*\hat{y}^* \in \partial f(\hat{x})$ *and* $\hat{y}^* \in \partial g(A\hat{x})$;
ii) $f(\hat{x}) + f^*(-A^*\hat{y}^*) = \langle -A^*\hat{y}^*, \hat{x} \rangle$ *and* $g(A\hat{x}) + g^*(\hat{y}^*) = \langle \hat{y}^*, A\hat{x} \rangle$;
iii) $f(\hat{x}) + g(A\hat{x}) + f^*(-A^*\hat{y}^*) + g^*(\hat{y}^*) = 0$; *and*
iv) $\hat{x} \in S$ *and* $\hat{y}^* \in S^*$ *and* $\alpha + \alpha^* = 0$.

Moreover, if $\hat{x} \in S$ and there is $x \in \mathrm{dom}(f)$ such that g is continuous in Ax, then there exists $\hat{y}^* \in Y^*$ such that all four statements hold.

Proof. Statements i) and ii) are equivalent by the Fenchel–Young Inequality (Proposition 3.42) and, clearly, they imply iii). But iii) implies ii) since

$$\left[f(x) + f^*(-A^*y^*) - \langle -A^*y^*, x \rangle \right] + \left[g(Ax) + g^*(y^*) - \langle y^*, Ax \rangle \right] = 0$$

and each term in brackets is nonnegative. Next, iii) and iv) are equivalent in view of (3.19). Finally, if $\hat{x} \in S$ and there is $x \in \mathrm{dom}(f)$ such that g is continuous in Ax, then

$$0 \in \partial(f + g \circ A)(\hat{x}) = \partial f(\hat{x}) + A^* \partial g(A\hat{x})$$

by the Chain Rule (Proposition 3.28) and the Moreau–Rockafellar Theorem 3.30. Hence, there is $\hat{y}^* \in \partial g(A\hat{x})$ such that $-A^*\hat{y}^* \in \partial f(\hat{x})$, which is i). □

For the *linear programming problem*, the preceding argument gives:

Example 3.52. Consider the *linear programming problem*:

$$(LP) \qquad\qquad \min_{x \in \mathbf{R}^N} \{c \cdot x : Ax \leq b\},$$

where $c \in \mathbf{R}^N$, A is a matrix of size $M \times N$, and $b \in \mathbf{R}^M$. This problem can be recast in the form of (P) by setting $f(x) = c \cdot x$ and $g(y) = \delta_{R_+^M}(b - y)$. The Fenchel conjugates are easily computed (Examples 3.46 and 3.47) and the dual problem is

$$(DLP) \qquad\qquad \min_{y^* \in \mathbf{R}^M} \{b \cdot y^* : A^*y^* + c = 0, \text{ and } y^* \geq 0\}.$$

If (LP) has a solution and there is $x \in \mathbf{R}^N$ such that $Ax < b$, then the dual problem has a solution, there is no duality gap and the primal-dual solutions $(\hat{x}, \hat{y}^*)$ are characterized by Theorem 3.51. Observe that the dimension of the space of variables, namely N, is often much larger than the number of constraints, which is M. In those cases, the dual problem (DLP) is much smaller than the primal one. □

3.6.3 The Biconjugate

If f^* is proper, the *biconjugate* of f is the function $f^{**} : X \to \mathbf{R} \cup \{+\infty\}$ defined by

$$f^{**}(x) = \sup_{x^* \in X^*} \{\langle x^*, x \rangle - f^*(x^*)\}.$$

Being a supremum of continuous affine functions, the biconjugate f^{**} is convex and lower-semicontinuous.

Remark 3.53. From the Fenchel–Young Inequality (Proposition 3.42), we deduce that $f^{**}(x) \leq f(x)$ for all $x \in X$. In particular, f^{**} is proper. A stronger conclusion for lower-semicontinuous convex functions is given in Proposition 3.56. □

Remark 3.54. We have defined the biconjugate in the original space X and not in the bidual X^{**}. Therefore, $f^{**} \neq (f^*)^*$ (although $f^{**}(x) = (f^*)^*(\mathscr{J}(x))$ for all $x \in X$, where $\mathscr{J}$ is the canonical embedding of X into X^{**}), but this difference disappears in reflexive spaces. □

Example 3.55. Let $f : X \to \mathbf{R}$ be defined by $f(x) = \langle x_0^*, x \rangle + \alpha$ for some $x_0^* \in X^*$ and $\alpha \in \mathbf{R}$. We already computed $f^* = \delta_{\{x_0^*\}} - \alpha$ in Example 3.46. We deduce that

$$f^{**}(x) = \sup_{x^* \in X^*} \{\langle x^*, x \rangle - f^*(x^*)\} = \langle x_0^*, x \rangle + \alpha.$$

Therefore, $f^{**} = f$. □

The situation discussed in Example 3.55 for continuous affine functions is not exceptional. On the contrary, we have the following:

Proposition 3.56. *Let* $f : X \to \mathbf{R} \cup \{+\infty\}$ *be proper. Then,* f *is convex and lower-semicontinuous if, and only if,* $f^{**} = f$.

Proof. We already mentioned in Remark 3.53 that always $f^{**} \leq f$. On the other hand, since f is convex and lower-semicontinuous, there exists a family $(f_i)_{i \in I}$ of continuous affine functions on X such that $f = \sup_{i \in I}(f_i)$, by Proposition 3.1. As in Proposition 3.50, we see that $f \leq g$ implies $f^{**} \leq g^{**}$. Therefore,

$$f^{**} \geq \sup_{i \in I}(f_i^{**}) = \sup_{i \in I}(f_i) = f,$$

because $f_i^{**} = f_i$ for continuous affine functions (see Example 3.55). The converse is straightforward, since $f = f^{**}$ is a supremum of continuous affine functions. $\square$

Corollary 3.57. *Let* $f : X \to \mathbf{R} \cup \{+\infty\}$ *be proper. Then,* f^{**} *is the greatest lower-semicontinuous and convex function below* f.

Proof. If $g : X \to \mathbf{R} \cup \{+\infty\}$ is a lower-semicontinuous and convex function such that $g \leq f$, then $g = g^{**} \leq f^{**}$. $\square$

Proposition 3.56 also gives the following converse to Proposition 3.20:

Proposition 3.58. *Let* $f : X \to \mathbf{R} \cup \{+\infty\}$ *be convex. Suppose* f *is continuous in* x_0 *and* $\partial f(x_0) = \{x_0^*\}$. *Then* f *is Gâteaux-differentiable in* x_0 *and* $\nabla f(x_0) = x_0^*$.

Proof. As we saw in Proposition 3.9, the function $\phi_{x_0} : X \to \mathbf{R}$ defined by $\phi_{x_0}(h) = f'(x_0; h)$ is convex and continuous in X. We shall see that actually $\phi_{x_0}(h) = \langle x_0^*, h \rangle$ for all $h \in X$. First observe that

$$\phi_{x_0}(h) = \phi_{x_0}^{**}(h) = \sup_{x^* \in X^*} \{ \langle x^*, h \rangle - \phi_{x_0}^*(x^*) \}.$$

Next, as we computed in Example 3.49, $\phi_{x_0}^*(x^*) = \delta_{\partial f(x_0)}(x^*)$. Hence,

$$\phi_{x_0}(h) = \sup_{x^* \in X^*} \{ \langle x^*, h \rangle - \delta_{\{x_0^*\}}(x^*) \} - \langle x_0^*, h \rangle.$$

We conclude that f is Gâteaux-differentiable in x_0 and $\nabla f(x_0) = x_0^*$. $\square$

Combining the Fenchel–Young Inequality (Proposition 3.42) and Proposition 3.56, we obtain the following:

Proposition 3.59 (Legendre–Fenchel Reciprocity Formula)**.** *Let* $f : X \to \mathbf{R} \cup \{+\infty\}$ *be proper, lower-semicontinuous, and convex. Then*

$$x^* \in \partial f(x) \qquad \textit{if, and only if,} \qquad \mathscr{J}(x) \in \partial f^*(x^*),$$

where $\mathscr{J}$ *is the canonical embedding of* X *into* X^{**}.

Remark 3.60. The dissymmetry introduced by defining ∂f^* as a subset of X^{**} and the biconjugate as a function of X may seem unpleasant. Actually, it does produce some differences, not only in Proposition 3.59, but, for instance, in the roles of primal and dual problems (see Sects. 3.6.2 and 3.7.2). Of course, all the dissymmetry disappears in reflexive spaces. In general Banach spaces, this difficulty may be circumvented by endowing X^* with the weak* topology σ^*. The advantage of doing so, is that the topological dual of X^* is X (see [94, Theorem 3.10]). In this setting, X^* will no longer be a normed space, but the core aspects of this chapter can be recast in the context of *locally convex topological vector spaces in duality*. □

3.7 Optimality Conditions for Constrained Problems

Let $C \subset X$ be closed and convex, and let $f : X \to \mathbf{R} \cup \{+\infty\}$ be proper, lower-semicontinuous, and convex. The following result characterizes the solutions of the optimization problem

$$\min\{\, f(x) : x \in C \,\}.$$

Proposition 3.61. *Let $f : X \to \mathbf{R} \cup \{+\infty\}$ be proper, lower-semicontinuous, and convex and let $C \subset X$ be closed and convex. Assume either that f is continuous at some point of C, or that there is an interior point of C where f is finite. Then $\hat{x}$ minimizes f on C if, and only if, there is $p \in \partial f(\hat{x})$ such that $-p \in N_C(\hat{x})$.*

Proof. By Fermat's Rule (Theorem 3.24) and the Moreau–Rockafellar Theorem 3.30, $\hat{x}$ minimizes f on C if, and only if,

$$0 \in \partial(f + \delta_C)(\hat{x}) = \partial f(\hat{x}) + N_C(\hat{x}).$$

This is equivalent to the existence of $p \in \partial f(\hat{x})$ such that $-p \in N_C(\hat{x})$. □

Let C be a closed affine subspace of X, namely, $C = \{x_0\} + V$, where $x_0 \in X$ and V is a closed subspace of X. Then $N_C(\hat{x}) = V^{\perp}$. We obtain:

Corollary 3.62. *Let $C = \{x_0\} + V$, where $x_0 \in X$ and V is a closed subspace of X, and let $f : X \to \mathbf{R} \cup \{+\infty\}$ be a proper, lower-semicontinuous, and convex function and assume that f is continuous at some point of C. Then $\hat{x}$ minimizes f on C if, and only if, $\partial f(\hat{x}) \cap V^{\perp} \neq \emptyset$.*

3.7.1 Affine Constraints

Let $A \in \mathscr{L}(X;Y)$ and let $b \in Y$. We shall derive optimality conditions for the problem of minimizing a function f over a set C of the form:

$$C = \{x \in X : Ax = b\}, \tag{3.20}$$

which we assume to be nonempty. Before doing so, let us recall that, given $A \in \mathscr{L}(X;Y)$, the adjoint of A is the operator $A^* : Y^* \to X^*$ defined by the identity

$$\langle A^*y^*, x \rangle_{X^*,X} = \langle y^*, Ax \rangle_{Y^*,Y},$$

for $x \in X$ and $y^* \in Y^*$. It is possible to prove (see, for instance, [30, Theorem 2.19]) that, if A has closed range, then $\ker(A)^\perp = R(A^*)$.

We have the following:

Theorem 3.63. *Let C be defined by (3.20), where A has closed range. Let $f : X \to \mathbf{R} \cup \{+\infty\}$ be a proper, lower-semicontinuous, and convex function and assume that f is continuous at some point of C. Then $\hat{x}$ minimizes f on C if, and only if, $A\hat{x} = b$ and there is $\hat{y}^* \in Y^*$ such that $-A^*\hat{y}^* \in \partial f(\hat{x})$.*

Proof. First observe that $N_C(\hat{x}) = \ker(A)^\perp$. Since A has closed range, $\ker(A)^\perp = R(A^*)$. We deduce that $-\hat{p} \in N_C(\hat{x}) = R(A^*)$ if, and only if, $\hat{p} = A^*\hat{y}^*$ for some $\hat{y}^* \in Y^*$, and conclude using Proposition 3.61. $\square$

3.7.2 Nonlinear Constraints and Lagrange Multipliers

Throughout this section, we consider proper, lower-semicontinuous, and convex functions $f, g_1, \ldots, g_m : X \to \mathbf{R} \cup \{+\infty\}$, along with continuous affine functions $h_1, \ldots, h_p : X \to \mathbf{R}$, which we assume to be linearly independent. We shall derive optimality conditions for the problem of minimizing f over the set C defined by

$$C = \{x \in X : g_i(x) \le 0 \text{ for all } i, \text{ and } h_j(x) = 0 \text{ for all } j\}, \qquad (3.21)$$

assuming this set is nonempty. Since each g_i is convex and each h_j is affine, the set C is convex. To simplify the notation, write

$$S = \operatorname{argmin}\{f(x) : x \in C\} \qquad \text{and} \qquad \alpha = \inf\{f(x) : x \in C\}.$$

We begin by showing the following intermediate result, which is interesting in its own right:

Proposition 3.64. *There exist $\lambda_0, \ldots, \lambda_m \ge 0$, and $\mu_1, \ldots \mu_p \in \mathbf{R}$ (not all zero) such that*

$$\lambda_0 \alpha \le \lambda_0 f(x) + \sum_{i=1}^{m} \lambda_i g_i(x) + \sum_{j=1}^{p} \mu_j h_j(x) \qquad (3.22)$$

for all $x \in X$.

Proof. If $\alpha = -\infty$, the result is trivial. Otherwise, set

$$A = \bigcup_{x \in X} \left\{ (f(x) - \alpha, +\infty) \times \left[\prod_{i=1}^{m} (g_i(x), +\infty) \right] \times \left[\prod_{j=1}^{p} \{h_j(x)\} \right] \right\}.$$

The set $A \subset \mathbf{R} \times \mathbf{R}^m \times \mathbf{R}^p$ is nonempty and convex. Moreover, by the definition of α, $0 \notin A$. Accodring to the finite-dimensional version of the Hahn-Banach Separation Theorem, namely Proposition 1.12, there exists $(\lambda_0, \ldots, \lambda_m, \mu_1, \ldots, \mu_p) \in \mathbf{R} \times \mathbf{R}^m \times \mathbf{R}^p \setminus \{0\}$ such that

$$\lambda_0 u_0 + \cdots + \lambda_m u_m + \mu_1 v_1 + \cdots + \mu_p v_p \geq 0 \qquad (3.23)$$

for all $(u_0, \ldots, u_m, v_1, \ldots, v_p) \in A$. If some $\lambda_i < 0$, the left-hand side of (3.23) can be made negative by taking u_i large enough. Therefore, $\lambda_i \geq 0$ for each i. Passing to the limit in (3.23), we obtain (3.22) for each $x \in \mathrm{dom}(f) \cap (\bigcap_{i=1}^m \mathrm{dom}(g_i))$. The inequality holds trivially in all other points. □

A more precise and useful result can be obtained under a *qualification condition*:

Slater's condition: There exists $x_0 \in \mathrm{dom}(f)$ such that $g_i(x_0) < 0$ for $i = 1, \ldots m$, and $h_j(x_0) = 0$ for $j = 1, \ldots p$.

Roughly speaking, this means that the constraint given by the system of inequalities is *thick* in the subspace determined by the affine equality constraints.

Corollary 3.65. *Assume Slater's condition holds. Then, there exist $\hat{\lambda}_1, \ldots, \hat{\lambda}_m \geq 0$, and $\hat{\mu}_1, \ldots, \hat{\mu}_p \in \mathbf{R}$, such that*

$$\alpha \leq f(x) + \sum_{i=1}^m \hat{\lambda}_i g_i(x) + \sum_{j=1}^p \hat{\mu}_j h_j(x)$$

for all $x \in X$.

Proof. If $\lambda_0 = 0$ in Proposition 3.64, and Slater's condition holds, then $\lambda_1 = \cdots = \lambda_m = 0$. It follows that a nontrivial linear combination of the affine functions is nonnegative on X. By linearity, this combination must be identically 0, which contradicts the linear independence. It suffices to divide the whole expression by $\lambda_0 > 0$ and rename the other variables. □

As a consequence of the preceding discussion we obtain the first-order optimality condition for the constrained problem, namely:

Theorem 3.66. *If $\hat{x} \in S$ and Slater's condition holds, then there exist $\hat{\lambda}_1, \ldots, \hat{\lambda}_m \geq 0$, and $\hat{\mu}_1, \ldots, \hat{\mu}_p \in \mathbf{R}$, such that $\hat{\lambda}_i g_i(\hat{x}) = 0$ for all $i = 1, \ldots, m$ and*

$$0 \in \partial \left(f + \sum_{i=1}^m \hat{\lambda}_i g_i \right)(\hat{x}) + \sum_{j=1}^p \hat{\mu}_j \nabla h_j(\hat{x}). \qquad (3.24)$$

Conversely, if $\hat{x} \in C$ and there exist $\hat{\lambda}_1, \ldots, \hat{\lambda}_m \geq 0$, and $\hat{\mu}_1, \ldots, \hat{\mu}_p \in \mathbf{R}$, such that $\hat{\lambda}_i g_i(\hat{x}) = 0$ for all $i = 1, \ldots, m$ and (3.24) holds, then $\hat{x} \in S$.

Proof. If $\hat{x} \in S$, the inequality in Corollary 3.65 is in fact an equality and we easily see that $\hat{\lambda}_i g_i(\hat{x}) = 0$ for all $i = 1, \ldots, m$. Fermat's Rule (Theorem 3.24) gives (3.24). Conversely, we have

$$f(\hat{x}) \leq f(x) + \sum_{i=1}^{m} \hat{\lambda}_i g_i(x) + \sum_{j=1}^{p} \hat{\mu}_j h_j(x)$$

$$\leq f(x) + \sup_{\lambda_i \geq 0} \sup_{\mu_j \in \mathbf{R}} \left[\sum_{i=1}^{m} \lambda_i g_i(x) + \sum_{j=1}^{p} \mu_j h_j(x) \right]$$

$$= f(x) + \delta_C(x)$$

for all $x \in X$. It follows that $\hat{x}$ minimizes f on C.   □

Remark 3.67. By the Moreau–Rockafellar Theorem 3.30, the inclusion

$$0 \in \partial f(\hat{x}) + \sum_{i=1}^{m} \hat{\lambda}_i \partial g_i(\hat{x}) + \sum_{j=1}^{p} \hat{\mu}_j \nabla h_j(\hat{x})$$

implies (3.24). Moreover, they are equivalent if $f, g_1, \ldots, g_m$ have a common point of continuity.   □

Lagrangian Duality

For $(x, \lambda, \mu) \in X \times \mathbf{R}_+^m \times \mathbf{R}^p$, the *Lagrangian* for the constrained optimization problem is defined as

$$L(x, \lambda, \mu) = f(x) + \sum_{i=1}^{m} \lambda_i g_i(x) + \sum_{j=1}^{p} \mu_j h_j(x).$$

Observe that

$$\sup_{(\lambda, \mu) \in \mathbf{R}_+^m \times \mathbf{R}^p} L(x, \lambda, \mu) = f(x) + \delta_C(x).$$

We shall refer to the problem of minimizing f over C as the *primal problem*. Its value is given by

$$\alpha = \inf_{x \in X} \left[\sup_{(\lambda, \mu) \in \mathbf{R}_+^m \times \mathbf{R}^p} L(x, \lambda, \mu) \right],$$

and S is called the set of *primal solutions*. By inverting the order in which the supremum and infimum are taken, we obtain the *dual problem*, whose value is

$$\alpha^* = \sup_{(\lambda, \mu) \in \mathbf{R}_+^m \times \mathbf{R}^p} \left[\inf_{x \in X} L(x, \lambda, \mu) \right].$$

The set S^* of points at which the supremum is attained is the set of *dual solutions* or *Lagrange multipliers.*

Clearly, $\alpha^* \leq \alpha$. The difference $\alpha - \alpha^*$ is the *duality gap* between the primal and the dual problems.

We say $(\hat{x}, \hat{\lambda}, \hat{\mu}) \in X \times \mathbf{R}^m_+ \times \mathbf{R}^p$ is a *saddle point* of L if

$$L(\hat{x}, \lambda, \mu) \leq L(\hat{x}, \hat{\lambda}, \hat{\mu}) \leq L(x, \hat{\lambda}, \hat{\mu}) \qquad (3.25)$$

for all $(x, \lambda, \mu) \in X \times \mathbf{R}^m_+ \times \mathbf{R}^p$.

Primal and dual solutions are further characterized by:

Theorem 3.68. *Let* $(\hat{x}, \hat{\lambda}, \hat{\mu}) \in X \times \mathbf{R}^m_+ \times \mathbf{R}^p$. *The following are equivalent:*

i) $\hat{x} \in S$, $(\hat{\lambda}, \hat{\mu}) \in S^*$ *and* $\alpha = \alpha^*$;
ii) $(\hat{x}, \hat{\lambda}, \hat{\mu})$ *is a saddle point of L; and*
iii) $\hat{x} \in C$, $\hat{\lambda}_i g_i(\hat{x}) = 0$ *for all* $i = 1, \ldots, m$ *and* (3.24) *holds.*

Moreover, if $S \neq \emptyset$ *and Slater's condition holds, then* $S^* \neq \emptyset$ *and* $\alpha = \alpha^*$.

Proof. Let i) hold. Since $\hat{x} \in S$, we have

$$L(\hat{x}, \lambda, \mu) \leq \sup_{\lambda, \mu} L(\hat{x}, \lambda, \mu) = \alpha$$

for all $(\lambda, \mu) \in \mathbf{R}^m_+ \times \mathbf{R}^p$. Next,

$$\alpha^* = \inf_x L(x, \hat{\lambda}, \hat{\mu}) \leq L(x, \hat{\lambda}, \hat{\mu}).$$

for all $x \in X$. Finally,

$$\alpha^* = \inf_x L(x, \hat{\lambda}, \hat{\mu}) \leq L(\hat{x}, \hat{\lambda}, \hat{\mu}) \leq \sup_{\lambda, \mu} L(\hat{x}, \lambda, \mu) = \alpha.$$

Since $\alpha^* = \alpha$, we obtain (3.25), which gives ii).

Now, suppose ii) holds. By (3.25), we have

$$f(\hat{x}) + \delta_C(\hat{x}) = \sup_{\lambda, \mu} L(\hat{x}, \lambda, \mu) \leq \inf_x L(x, \hat{\lambda}, \hat{\mu}) = \alpha^* < +\infty,$$

and so, $\hat{x} \in C$. Moreover,

$$f(\hat{x}) = f(\hat{x}) + \delta_C(\hat{x}) \leq L(\hat{x}, \hat{\lambda}, \hat{\mu}) = f(\hat{x}) + \sum_{i=1}^m \hat{\lambda}_i g_i(\hat{x}) \leq f(\hat{x}).$$

It follows that $\sum_{i=1}^m \hat{\lambda}_i g_i(\hat{x}) = 0$ and we conclude that each term is zero, since they all have the same sign. The second inequality in (3.25) gives (3.24).

Next, assume iii) holds. By Theorem 3.66, $\hat{x} \in S$. On the other hand,

$$L(\hat{x},\hat{\lambda},\hat{\mu}) = f(\hat{x}) = \alpha \geq \alpha^* = \sup_{\lambda,\mu}\left[\inf_{x} L(x,\lambda,\mu)\right],$$

and $(\hat{\lambda},\hat{\mu}) \in S^*$. We conclude that $\alpha^* = \inf_x L(x,\hat{\lambda},\hat{\mu}) = L(\hat{x},\hat{\lambda},\hat{\mu}) = \alpha$.

Finally, if $S \neq \emptyset$ and Slater's condition holds, by Theorem 3.66, there is $(\hat{\lambda},\hat{\mu})$ such that condition iii) in Theorem 3.68 holds. Then, $(\hat{\lambda},\hat{\mu}) \in S^*$ and $\alpha = \alpha^*$. $\square$

Remark 3.69. As we pointed out in Example 3.52 for the linear programming problem, the dimension of the space in which the dual problem is stated is the number of constraints. Therefore, the dual problem is typically easier to solve than the primal one. Moreover, according to Theorem 3.68, once we have found a Lagrange multiplier, we can recover the solution $\hat{x}$ for the original (primal) constrained problem as a solution of

$$\min_{x \in X} L(x,\hat{\lambda},\hat{\mu}),$$

which is an unconstrained problem. $\square$

Remark 3.70. In the setting of Theorem 3.63, we know that $\hat{x}$ minimizes f on $C = \{x \in X : Ax = b\}$ if, and only if, $A\hat{x} = b$ and there is $\hat{y}^* \in Y^*$ such that $-A^*\hat{y}^* \in \partial f(\hat{x})$. If we define the Lagrangian $L : X \times Y^* \to \mathbf{R} \cup \{+\infty\}$ by

$$L(x,y^*) = f(x) + \langle y^*, Ax - b \rangle,$$

then, the following are equivalent:

i) $A\hat{x} = b$ and $-A^*\hat{y}^* \in \partial f(\hat{x})$;
ii) $(\hat{x},\hat{y}^*)$ is a saddle point of L; and
iii) $\hat{x} \in C$, and minimizes $x \mapsto L(x,\hat{y}^*)$ over X.

This allows us to characterize $\hat{x}$ as a solution for an unconstrained problem. In this context, $\hat{y}^*$ is a (possibly infinite-dimensional) Lagrange multiplier. $\square$

Chapter 4
Examples

Abstract The tools presented in the previous chapters are useful, on the one hand, to prove that a wide variety of optimization problems have solutions; and, on the other, to provide useful characterizations allowing to determine them. In this chapter, we present a short selection of problems to illustrate some of those tools. We begin by revisiting some results from functional analysis concerning the maximization of bounded linear functionals and the realization of the dual norm. Next, we discuss some problems in optimal control and calculus of variations. Another standard application of these convex analysis techniques lies in the field of elliptic partial differential equations. We shall review the theorems of Stampacchia and Lax-Milgram, along with some variations of Poisson's equation, including the obstacle problem and the p-Laplacian. We finish by commenting a problem of data compression and restoration.

4.1 Norm of a Bounded Linear Functional

Let X be a normed space and take $L \in X^*$. Recall that $\|L\|_* = \sup_{\|x\|=1} |L(x)|$. By linearity, it is easy to prove that

$$\sup_{\|x\|=1} |L(x)| = \sup_{\|x\|<1} |L(x)| = \sup_{\|x\|\leq 1} |L(x)|.$$

We shall focus in this section on the third form, since it can be seen as a convex minimization problem. Indeed, the value of

$$\inf_{x \in X} \left\{ L(x) + \delta_{\bar{B}(0,1)}(x) \right\}$$

is precisely $-\|L\|_*$.

Let B be any nonempty, closed, convex, and bounded subset of X. The function $f = L + \delta_B$ is proper, lower-semicontinuous, convex, and coercive. If X is reflexive, the infimum is attained, by Theorem 2.19. We obtain the following:

Corollary 4.1. *Let X be reflexive and let $L \in X^*$. Then, L attains its maximum and its minimum on each nonempty, closed, convex, and bounded subset of X. In particular, there exists $x_0 \in \bar{B}(0,1)$ such that $L(x_0) = \|L\|_*$.*

The last part can also be seen as a consequence of Corollary 1.16. Actually, the converse is true as well: if every $L \in X^*$ attains its maximum in $\bar{B}(0,1)$, then X is reflexive (see [68]). Therefore, in nonreflexive spaces there are bounded linear functionals that do not realize their supremum on the ball. Here is an example:

Example 4.2. Consider (as in Example 1.2) the space $X = \mathscr{C}([-1,1];\mathbf{R})$ of continuous real-valued functions on $[-1,1]$, with the norm $\|\cdot\|_\infty$ defined by $\|x\|_\infty = \max_{t \in [-1,1]} |x(t)|$. Define $L : X \to \mathbf{R}$ by

$$L(x) = \int_{-1}^0 x(t)\,dt - \int_0^1 x(t)\,dt.$$

Clearly, $L \in X^*$ with $\|L\|_* \le 2$, since

$$|L(x)| \le \int_{-1}^0 \|x\|_\infty \, dt + \int_0^1 \|x\|_\infty \, dt = 2\|x\|_\infty.$$

We shall see that $\|L\|_* = 2$. Indeed, consider the sequence $(x_n)_{n \ge 2}$ in X defined by

$$x_n(t) = \begin{cases} 1 & \text{if } -1 \le t \le -\frac{1}{n} \\ -nt & \text{if } -\frac{1}{n} < t < \frac{1}{n} \\ -1 & \text{if } \frac{1}{n} \le t \le 1. \end{cases}$$

We easily see that $x_n \in \bar{B}(0,1)$ for all $n \ge 2$ and

$$\lim_{n \to \infty} L(x_n) = \lim_{n \to \infty} \left(2 - \frac{1}{n}\right) = 2.$$

However, if $\|x\|_\infty \le 1$ and $L(x) = 2$, then, necessarily, $x(t) = 1$ for all $t \in [-1,0]$ and $x(t) = -1$ for all $t \in [0,1]$. Therefore, x cannot belong to X. $\square$

As a consequence of Corollary 4.1 and Example 4.2, we obtain:

Corollary 4.3. *The space $\left(\mathscr{C}([-1,1];\mathbf{R}), \|\cdot\|_\infty\right)$ is not reflexive.*

As we have pointed out, if X is not reflexive, then not all the elements of X^* attain their supremum on the unit ball. However, *many* of them do. More precisely, we have the following remarkable result of Functional Analysis:

Corollary 4.4 (Bishop-Phelps Theorem). *Let X be a Banach space and let $B \subset X$ be nonempty, closed, convex, and bounded. Then, the set of all bounded linear functionals that attain their maximum in B is dense in X^*.*

Proof. As usual, let δ_B be the indicator function of the set B. Then

$$\delta_B^*(L) = \sup_{x\in B}\langle L,x\rangle$$

for each $L \in X^*$ (see also Example 3.47). Since B is bounded, $\mathrm{dom}(\delta_B^*) = X^*$. On the other hand, L attains its maximum in B if, and only if, $L \in R(\partial \delta_B)$ by Fermat's Rule (Theorem 3.24) and the Moreau–Rockafellar Theorem 3.30. But, by Proposition 3.43, $\overline{R(\partial \delta_B)} = \overline{\mathrm{dom}(\delta_B^*)} = X^*$. $\qquad\square$

4.2 Optimal Control and Calculus of Variations

Using convex analysis and differential calculus tools, it is possible to prove the existence and provide characterizations for the solutions of certain problems in optimal control and calculus of variations.

4.2.1 Controlled Systems

Let $y_0 \in \mathbf{R}^n$, $A : [0,T] \to \mathbf{R}^{n\times n}$, $B : [0,T] \to \mathbf{R}^{n\times m}$, and $c : [0,T] \to \mathbf{R}^n$. Consider the *linear control system*

$$(CS) \qquad \begin{cases} \dot{y}(t) = A(t)y(t) + B(t)u(t) + c(t), & t \in (0,T) \\ y(0) = y_0. \end{cases}$$

We assume, for simplicity, that the functions A, B, and c are continuous.

Given $u \in L^p(0,T;\mathbf{R}^M)$ with $p \in [1,\infty]$, one can find $y_u \in \mathscr{C}(0,T;\mathbf{R}^N)$ such that the pair (u,y_u) satisfies (CS). Indeed, let $R : [0,T] \to \mathbf{R}^{n\times n}$ be the resolvent of the matrix equation $\dot{X} = AX$ with initial condition $X(0) = I$. Then y_u can be computed by using the Variation of Parameters Formula:

$$y_u(t) = R(t)y_0 + R(t)\int_0^t R(s)^{-1}[B(s)u(s) + c(s)]\,ds. \qquad (4.1)$$

It is easy to see that the function $u \mapsto y_u$ is affine and continuous.

The system (CS) is *controllable* from y_0 to the *target set* $\mathscr{T} \subset \mathbf{R}^N$ in time T if there exists $u \in L^p(0,T;\mathbf{R}^M)$ such that $y_u(T) \in \mathscr{T}$. In other words, if the function $\Phi : L^p(0,T;\mathbf{R}^M) \to \mathbf{R} \cup \{+\infty\}$, defined by $\Phi(u) = \delta_{\mathscr{T}}(y_u(T))$, is proper.

4.2.2 Existence of an Optimal Control

Consider the functional J defined on some function space X (to be specified later) by

$$J[u] = \int_0^T \ell(t, u(t), y_u(t)) \, dt + h(y_u(T))$$

with $\ell : \mathbf{R} \times \mathbf{R}^M \times \mathbf{R}^N \to \mathbf{R} \cup \{+\infty\}$. The *optimal control problem* is

(OC) $\qquad\qquad\qquad\qquad \min\{J[u] : u \in X\}.$

It is *feasible* if J is proper. In this section we focus our attention on the case where the function ℓ has the form

$$\ell(t, v, x) = f(v) + g(x). \tag{4.2}$$

We assume $f : \mathbf{R}^M \to \mathbf{R} \cup \{+\infty\}$ is proper, convex, and lower-semicontinuous, and that there exist $\alpha_f \in \mathbf{R}$ and $\beta > 0$ such that

(G) $\qquad\qquad\qquad\qquad f(v) \geq \alpha_f + \beta |v|^p$

for all $v \in \mathbf{R}^M$ and some $p \in (1, \infty)$. This p will determine the function space where we shall search for solutions. Let $g : \mathbf{R}^N \to \mathbf{R}$ be continuous and bounded from below by α_g, and let $h : \mathbf{R}^N \to \mathbf{R} \cup \{+\infty\}$ be lower-semicontinuous (for the strong topology) and bounded from below by α_h. We shall prove that J has a minimizer.

Theorem 4.5. *If (OC) is feasible, it has a solution in $L^p(0, T; \mathbf{R}^M)$.*

Proof. By Theorem 2.9, it suffices to verify that J is sequentially inf-compact and sequentially lower-semicontinuous for the weak topology.

For inf-compactness, since $L^p(0, T; \mathbf{R}^M)$ is reflexive, it suffices to show that the sublevel sets are bounded, by Theorem 1.24. Fix $\gamma \in \mathbf{R}$ and take $u \in \Gamma_\gamma(J)$. By the growth condition (G), we have

$$f(u(t)) \geq \alpha_f + \beta |u(t)|^p$$

for almost every $t \in [0, T]$. Integrating from 0 to T we obtain

$$\gamma \geq J[u] \geq (\alpha_f + \alpha_g + \alpha_h)T + \beta \|u\|_{L^p(0,T;\mathbf{R}^M)}^p.$$

Setting $R = \beta^{-1}[\gamma - (\alpha_f + \alpha_g + \alpha_h)T]$, we have $\Gamma_\gamma(J) \subset B(0, \sqrt{R})$.

For the lower-semicontinuity, let (u_n) be a sequence in H such that $u_n \rightharpoonup \bar{u}$ as $n \to \infty$. We shall prove that

$$J[\bar{u}] \leq \liminf_{n \to \infty} J[u_n]. \tag{4.3}$$

To simplify the notation, for $t \in [0,T]$, set

$$z_0(t) = R(t)y_0 + R(t) \int_0^t R(s)^{-1}c(s)\, dt,$$

which is the part of y_u that does not depend on u. The j-th component of y_{u_n} satisfies

$$y_{u_n}^j(t) = z_0^j(t) + \int_0^T 1_{[0,t]}(s)\mu_j(t,s)u_n(s)\, dt,$$

where $\mu_j(t,s)$ is the j-th row of the matrix $R(t)R(s)^{-1}B(s)$. The weak convergence of (u_n) to $\bar{u}$ implies

$$\lim_{n\to\infty} y_{u_n}(t) = z_0(t) + R(t) \int_0^t R(s)^{-1}B(s)\bar{u}(s)\, dt = y_{\bar{u}}(t)$$

for each $t \in [0,T]$. Moreover, this convergence is uniform because the sequence (y_{u_n}) is equicontinuous. Indeed, if $0 \le \tau \le t \le T$, we have

$$y_{u_n}^j(t) - y_{u_n}^j(\tau) = \int_0^T 1_{[\tau,t]}(s)\mu_j(t,s)u_n(s)\, dt.$$

But $\mu_j(t,s)$ is bounded by continuity, while the sequence (u_n) is bounded because it is weakly convergent. Therefore,

$$|y_{u_n}^j(t) - y_{u_n}^j(\tau)| \le C|t - \tau|$$

for some constant C, independent of n. It immediately follows that

$$\lim_{n\to\infty} \int_0^T g(y_{u_n}(t))\, dt = \int_0^T g(y_{\bar{u}}(t))\, dt$$

and

$$\liminf_{n\to\infty} h(y_{u_n}(T)) \ge h(y_{\bar{u}}(T)).$$

Therefore, in order to establish (4.3), it suffices to prove that

$$\liminf_{n\to\infty} \int_0^T f(u_n(t))\, dt \ge \int_0^T f(\bar{u}(t))\, dt,$$

which was already done in Example 2.18. $\qquad \Box$

Remark 4.6. Requiring a controllability condition $y_u(T) \in \mathscr{T}$ is equivalent to adding a term $\delta_{\mathscr{T}}(y_u(T))$ in the definition of $h(y_u(T))$. $\qquad \Box$

4.2.3 The Linear-Quadratic Problem

We shall derive a characterization for the solutions of the *linear-quadratic control problem*, where the objective function is given by

$$J[u] = \frac{1}{2} \int_0^T u(t)^* U(t) u(t)\, dt + \frac{1}{2} \int_0^T y_u(t)^* W(t) y_u(t)\, dt + h(y_u(T)),$$

where

- The vector functions u and y_u are related by (CS) but, for simplicity, we assume $c \equiv 0$.
- All vectors are represented as columns, and the star $*$ denotes the transposition of vectors and matrices.
- For each t, $U(t)$ is a uniformly elliptic symmetric matrix of size $M \times M$, and the function $t \mapsto U(t)$ is continuous. With this, there is a constant $\alpha > 0$ such that $u^* U(t) u \geq \alpha \|u\|^2$ for all $u \in \mathbf{R}^M$ and $t \in [0, T]$.
- For each t, $W(t)$ is a positive semidefinite symmetric matrix of size $N \times N$, and the function $t \mapsto W(t)$ is continuous.
- The function h is convex and differentiable. Its gradient ∇h is represented as a column vector.

By applying Theorem 4.5 in $L^2(0, T; \mathbf{R}^M)$, the optimal control problem has a solution. Moreover, it is easy to see that J is strictly convex, and so, there is a unique optimal pair $(\bar{u}, \bar{y}_u)$. We have the following:

Theorem 4.7. *The pair $(\bar{u}, \bar{y}_u)$ is optimal if, and only if,*

$$\bar{u}(t) = U(t)^{-1} B(t)^* p(t) \tag{4.4}$$

for almost every $t \in [0, T]$, where the adjoint state p *is the unique solution of the* adjoint equation
$$\dot{p}(t) = -A(t)^* p(t) + W(t) \bar{y}_u(t)$$
with terminal condition $p(T) = -\nabla h(\bar{y}_u(T))$.

Proof. Since J is convex and differentiable, by Fermat's Rule (Theorem 1.32), we know that $(\bar{u}, \bar{y}_u)$ is optimal if, and only if,

$$0 = \langle \nabla J(\bar{u}), v \rangle$$
$$= \int_0^T \bar{u}(t)^* U(t) v(t)\, dt + \int_0^T \bar{y}_u(t)^* W(t) y_v(t)\, dt + \nabla h(\bar{y}_u(T))^* y_v(T)$$

for all $v \in L^2(0, T; \mathbf{R}^M)$. Let us focus on the middle term first. By using the adjoint equation, we see that

$$\bar{y}_u(t)^* W(t) = \dot{p}(t)^* + p(t)^* A(t).$$

Therefore,

$$\int_0^T \bar{y}_u(t)^* W(t) y_v(t)\,dt = \int_0^T \dot{p}(t)^* y_v(t)\,dt + \int_0^T p(t)^* A(t) y_v(t)\,dt$$

$$= p(T)^* y_v(T) - \int_0^T p(t)^* \left[\dot{y}_v(t) - A(t) y_v(t) \right] dt$$

$$= -\nabla h(\bar{y}_u(T))^* y_v(T) - \int_0^T p(t)^* B(t) v(t)\,dt.$$

We conclude that

$$0 = \langle \nabla J(\bar{u}), v \rangle = \int_0^T \left[\bar{u}(t)^* U(t) - p(t)^* B(t) \right] v(t)\,dt$$

for all $v \in L^2(0,T;\mathbf{R}^M)$, and the result follows. $\Box$

Let us analyze the *rocket-car* problem:

Example 4.8. A car of unit mass is equipped with propulsors on each side, to move it along a straight line. It starts at the point x_0 with initial velocity v_0. The control u is the force exerted by the propulsors, so that the position x and velocity v satisfy:

$$\begin{pmatrix} \dot{x}(t) \\ \dot{v}(t) \end{pmatrix} = \begin{pmatrix} 0 & 1 \\ 0 & 0 \end{pmatrix} \begin{pmatrix} x(t) \\ v(t) \end{pmatrix} + \begin{pmatrix} 0 \\ 1 \end{pmatrix} u(t), \qquad \begin{pmatrix} x(0) \\ v(0) \end{pmatrix} = \begin{pmatrix} x_0 \\ v_0 \end{pmatrix}.$$

Consider the optimal control problem of minimizing the functional

$$J[u] = x_u(T) + \frac{1}{2} \int_0^T u(t)^2\,dt,$$

which we can interpret as moving the car far away to the left, without spending too much energy. The adjoint equation is

$$\begin{pmatrix} \dot{p}_1(t) \\ \dot{p}_2(t) \end{pmatrix} = -\begin{pmatrix} 0 & 0 \\ 1 & 0 \end{pmatrix} \begin{pmatrix} p_1(t) \\ p_2(t) \end{pmatrix}, \qquad \begin{pmatrix} p_1(T) \\ p_2(T) \end{pmatrix} = \begin{pmatrix} -1 \\ 0 \end{pmatrix},$$

from which we deduce that $p_1(t) \equiv -1$ and $p_2(t) = t - T$. Finally, from (4.4), we deduce that $\bar{u}(t) = p_2(t) = t - T$. With this information and the initial condition, we can compute $\bar{x}(T)$ and the value $J[\bar{u}]$. $\Box$

Introducing Controllability

Assume now that $h = \delta_{\mathscr{T}}$, which means that we require that the terminal state $\bar{y}_u(T)$ belong to a convex target set $\mathscr{T}$ (see Remark 4.6). Although h is no longer differentiable, one can follow pretty much the same argument as in the proof of Theorem 4.7 (but using the nonsmooth version of Fermat's Rule, namely 3.24) to deduce that the adjoint state p must satisfy the terminal condition $-p(T) \in N_{\mathscr{T}}(\bar{y}_u(T))$.

Example 4.9. If the distance between the terminal state and some reference point y_T must be less than or equal to $\rho > 0$, then $\mathscr{T} = \bar{B}(y_T, \rho)$. There are two possibilities for the terminal states:

1. either $\|\bar{y}_u(T) - y_T\| < \rho$ and $p(T) = 0$;
2. or $\|\bar{y}_u(T) - y_T\| = \rho$ and $p(T) = \kappa(y_T - \bar{y}_u(T))$ for some $\kappa \geq 0$. $\qquad\square$

4.2.4 Calculus of Variations

The classical problem of Calculus of Variations is

$$(CV) \qquad \min\{ J[x] \ : \ x \in AC(0,T;\mathbf{R}^N), \ x(0) = x_0, \ x(T) = x_T \},$$

where the function J is of the form

$$J[x] = \int_0^T \ell(t, \dot{x}(t), x(t)) \, dt$$

for some function $\ell : \mathbf{R} \times \mathbf{R}^N \times \mathbf{R}^N \to \mathbf{R}$. If ℓ is of the form described in (4.2), namely

$$\ell(t, v, x) = f(v) + g(x),$$

this problem fits in the framework of (OC) by setting $M = N$, $A \equiv 0$, $B = I$, $c \equiv 0$ and $h = \delta_{\{x_T\}}$. From Theorem 4.5, it ensues that (CV) has a solution whenever it is feasible.

Optimality Condition: Euler–Lagrange Equation

If ℓ is of class $\mathscr{C}^1$, the function J is differentiable (see Example 1.28) and

$$DJ(x)h = \int_0^T \left[\nabla_2 \ell(t, x(t), \dot{x}(t)) \cdot h(t) + \nabla_3 \ell(t, x(t), \dot{x}(t)) \cdot \dot{h}(t) \right] dt,$$

where ∇_i denotes the gradient with respect to the i-th set of variables.

In order to obtain necessary optimality conditions, we shall use the following auxiliary result:

Lemma 4.10. *Let* $\alpha, \beta \in \mathscr{C}^0\left([0,T];\mathbf{R}\right)$ *be such that*

$$\int_0^T \left[\alpha(t)h(t)dt + \beta(t)\dot{h}(t) \right] dt = 0$$

for each $h \in \mathscr{C}^1\left([0,T];\mathbf{R}\right)$ *satisfying* $h(0) = h(T) = 0$. *Then* β *is continuously differentiable and* $\dot{\beta} = \alpha$.

Proof. Let us first analyze the case $\alpha \equiv 0$. We must prove that β is constant. To this end, define the function $H : [0, T] \to \mathbf{R}$ as

$$H(t) = \int_0^t [\beta(s) - B]\, ds, \qquad \text{where} \qquad B = \frac{1}{T} \int_0^T \beta(t)\, dt.$$

Since H is continuously differentiable and $H(0) = H(T) = 0$,

$$
\begin{aligned}
0 &= \int_0^T \beta(t)\dot{H}(t)\, dt \\
&= \int_0^T \beta(t)[\beta(t) - B]\, dt \\
&= \int_0^T [\beta(t) - B]^2\, dt + B \int_0^T [\beta(t) - B]\, dt \\
&= \int_0^T [\beta(t) - B]^2\, dt.
\end{aligned}
$$

This implies $\beta \equiv B$ because β is continuous. For the general case, define

$$A(t) = \int_0^t \alpha(s)\, ds.$$

Using integration by parts, we see that

$$0 = \int_0^T \left[\alpha(t)h(t)dt + \beta(t)\dot{h}(t) \right] dt = \int_0^T \left[\beta(t) - A(t) \right] \dot{h}(t)\, dt.$$

for each $h \in \mathscr{C}^1\left([0,T]; \mathbf{R}\right)$ such that $h(0) = h(T) = 0$. As we have seen, this implies $\beta - A$ must be constant. In other words, β is a primitive of α. It follows that β is continuously differentiable and $\dot{\beta} = \alpha$. $\qquad \square$

We are now in position to present the necessary optimality condition for (CV):

Theorem 4.11 (Euler–Lagrange Equation). *Let x^* be a smooth solution of (CV). Then, the function $t \mapsto \nabla_3 L(t, x^*(t), \dot{x}^*(t))$ is continuously differentiable and*

$$\frac{d}{dt}\left[\nabla_3 L(t, x^*(t), \dot{x}^*(t)) \right] = \nabla_2 L(t, x^*(t), \dot{x}^*(t))$$

for every $t \in (0, T)$.

Proof. Set

$$C_0 = \left\{ x \in \mathscr{C}^1\left([0, T]; \mathbf{R}^N\right) : x(0) = x(T) = 0 \right\},$$

and define $g : C_0 \to \mathbf{R}$ as

$$g(h) = f(x^* + h).$$

Clearly, 0 minimizes g on C_0 because x^* is a smooth solution of *(CV)*. In other words, $0 \in \operatorname{argmin}_{C_0}(g)$. Moreover, $Dg(0) = Df(x^*)$ and so, Fermat's Rule (Theorem 1.32) gives $Df(x^*)h = Dg(0)h = 0$ for every $h \in C_0$. This is precisely

$$\int_0^T \left[\nabla_2 L(t,x(t),\dot{x}(t)) \cdot h(t) + \nabla_3 L(t,x(t),\dot{x}(t)) \cdot \dot{h}(t) \right] dt = 0$$

for each $h \in C_0$. To conclude, for $k = 1,\ldots,N$ set $h_j \equiv 0$ for $j \neq k$, write $\alpha_k(t) = \left(\nabla_2 L(t,x^*(t),\dot{x}^*(t)) \right)_k$ and use Lemma 4.10 componentwise. $\square$

4.3 Some Elliptic Partial Differential Equations

The theory of partial differential equations makes systematic use of several tools that lie in the interface between functional and convex analysis. Here, we comment a few examples.

4.3.1 The Theorems of Stampacchia and Lax-Milgram

Theorem 4.12 (Stampacchia's Theorem). *Let X be reflexive and let $C \subset X$ be nonempty, closed, and convex. Given a continuous and symmetric bilinear form $B : X \times X \to \mathbf{R}$ and some $x^* \in X^*$, define $f : X \to \mathbf{R}$ by*

$$f(x) = \frac{1}{2}B(x,x) - \langle x^*, x \rangle,$$

and set $S = \mathrm{argmin}_C(f)$. We have the following:

i) If B is positive semidefinite and C is bounded, then $S \neq \emptyset$;
ii) If either B is positive definite and C is bounded, or B is uniformly elliptic, then S is a singleton $S = \{\bar{x}\}$.

In any case, $\bar{x} \in S$ if, and only if, $\bar{x} \in C$ and $B(\bar{x}, c - \bar{x}) \geq \langle x^*, c - \bar{x} \rangle$ for all $c \in C$.

Proof. The function $g = f + \delta_C$ is proper, lower-semicontinuous, convex, and coercive in both cases, and strictly convex in case ii), as shown in Example 2.11. The existence and uniqueness of minimizers follow from Theorem 2.19. On the other hand, by Fermat's Rule (Theorem 3.24) and the Moreau–Rockafellar Theorem 3.30 (see also Proposition 3.61 for a shortcut), $\bar{x} \in S$ if, and only if,

$$0 \in \partial g(\bar{x}) = \partial (f + \delta_C)(\bar{x}) = \partial f(\bar{x}) + \partial \delta_C(\bar{x}).$$

In other words, $-\nabla f(\bar{x}) \in N_C(\bar{x})$, which is equivalent to $B(\bar{x}, c - \bar{x}) \geq \langle x^*, c - \bar{x} \rangle$ for all $c \in C$ (see Example 1.27). $\square$

If $C = X$ we obtain the following:

Corollary 4.13 (Lax-Milgram Theorem). *Let X be reflexive, let $B : X \times X \to \mathbf{R}$ be a uniformly elliptic, continuous and symmetric bilinear function, and let $x^* \in X^*$. Then, the function $f : X \to \mathbf{R}$, defined by*

$$f(x) = \frac{1}{2}B(x,x) - \langle x^*, x \rangle,$$

has a unique minimizer $\bar{x}$. Moreover, $\bar{x}$ is the unique element of X that satisfies $B(\cdot, h) = \langle x^*, h \rangle$ for all $h \in X$.

4.3.2 Sobolev Spaces

We shall provide a brief summary of some Sobolev spaces, which are particularly useful for solving second-order elliptic partial differential equations. For more details, the reader may consult [1, 30] or [54].

Let Ω be a nonempty open subset of $\mathbf{R}^N$ and let $p \geq 1$. Consider the space

$$W^{1,p} = \{u \in L^p : \nabla u \in (L^p)^N\}$$

with the norm

$$\|u\|_{W^{1,p}} = \left[\int_\Omega \left[|\nabla u(\xi)|^p + |u(\xi)|^p \right] d\xi \right]^{1/p}.$$

The gradient is taken in the sense of distributions. We omit the reference to the domain Ω to simplify the notation, since no confusion should arise.

The space $W_0^{1,p}$ is the closure in $W^{1,p}$ of the subspace of infinitely differentiable functions in Ω whose support is a compact subset of Ω. Intuitively, it consists of the functions that vanish on the boundary $\partial\Omega$ of Ω. The following property will be useful later:

Proposition 4.14 (Poincaré's Inequalities). *Let Ω be bounded.*

i) There exists $C > 0$ such that

$$\|u\|_{L^p} \leq C\|\nabla u\|_{(L^p)^N}$$

for all $u \in W_0^{1,p}$. In particular, the function $u \mapsto \|\nabla u\|_{(L^p)^N}$ is a norm in $W_0^{1,p}$, which is equivalent to the original norm.

ii) If Ω is bounded, connected and $\partial\Omega$ is smooth, there exists $C > 0$ such that

$$\|u - \bar{u}\|_{L^p} \leq C\|\nabla u\|_{(L^p)^N}$$

for all $u \in W^{1,p}$. Here, we have written $\bar{u} = \frac{1}{|\Omega|} \int_\Omega u(\xi) d\xi$, where $|\Omega|$ is the measure of Ω.

If $p > 1$, then $W^{1,p}$ and $W_0^{1,p}$ are reflexive. We shall restrict ourselves to this case. We have

$$W_0^{1,p} \subset W^{1,p} \subset L^p$$

with continuous embeddings. The topological duals satisfy

$$L^q = (L^p)^* \subset (W^{1,p})^* \subset (W_0^{1,p})^*,$$

where $q > 1$ is defined by $\frac{1}{p} + \frac{1}{q} = 1$. Now, for each $v^* \in (W_0^{1,p})^*$, there exist $v_0, v_1, \ldots, v_N \in L^q$ such that

$$\langle v^*, u \rangle_{(W_0^{1,p})^*, W_0^{1,p}} = \int_\Omega v_0(\xi) u(\xi) \, d\xi + \sum_{i=1}^N \int_\Omega v_i(\xi) \frac{\partial u}{\partial x_i}(\xi) \, d\xi$$

for all $u \in W_0^{1,p}$. In turn, given $v_0, v_1, \ldots, v_N \in L^q$, we can define an element v^* of $(W^{1,p})^*$ by

$$\langle v^*, u \rangle_{(W^{1,p})^*, W^{1,p}} = \int_\Omega v_0(\xi) u(\xi) \, d\xi + \sum_{i=1}^N \int_\Omega v_i(\xi) \frac{\partial u}{\partial x_i}(\xi) \, d\xi$$

for all $u \in W^{1,p}$.

The case $p = 2$, namely $H^1 := W^{1,2}$, is of particular importance. It is a Hilbert space, whose norm is associated to the inner product

$$\langle u, v \rangle_{H^1} = \langle u, v \rangle_{L^2} + \langle \nabla u, \nabla v \rangle_{(L^2)^N}.$$

We also write $H_0^1 := W_0^{1,2}$. It is sometimes convenient not to identify H^1 and H_0^1 with their duals. Instead, one can see them as subspaces of L^2, identify the latter with itself and write

$$H_0^1 \subset H^1 \subset L^2 = (L^2)^* \subset (H^1)^* \subset (H_0^1)^*.$$

4.3.3 Poisson-Type Equations in H^1 and $W^{1,p}$

Dirichlet Boundary Conditions

Let $\Omega \subset \mathbf{R}^N$ be bounded. Given $\mu \in \mathbf{R}$, consider the bilinear function $B : H_0^1 \times H_0^1 \to \mathbf{R}$ defined by

$$B(u, v) = \mu \langle u, v \rangle_{L^2} + \langle \nabla u, \nabla v \rangle_{(L^2)^N}.$$

Clearly, B is continuous and symmetric. It is not difficult to see that it is uniformly elliptic if $\mu > -1/C^2$, where C is given by Poincaré's Inequality (part i) of Proposition 4.14).

Now take $v^* \in (H_0^1)^*$. By the Lax-Milgram Theorem (Corollary 4.13), there is a unique $\bar{u} \in H_0^1$ such that

$$\mu \langle \bar{u}, v \rangle_{L^2} + \langle \nabla \bar{u}, \nabla v \rangle_{(L^2)^N} = \langle v^*, v \rangle_{(H_0^1)^*, H_0^1}$$

for all $v \in H_0^1$. In other words, there is a unique weak solution $\bar{u}$ for the homogeneous Dirichlet boundary condition problem

$$\begin{cases} -\Delta u + \mu u = v^* \text{ in } \Omega \\ \qquad\quad u = 0 \text{ in } \partial\Omega. \end{cases}$$

Moreover, the function $\Phi : H_0^1 \to (H_0^1)^*$ defined by

$$\langle \Phi(u), v \rangle_{(H_0^1)^*, H_0^1} = B(u, v)$$

is a continuous isomorphism with continuous inverse.

Suppose now that $g \in H^1$ is uniformly continuous on Ω (and extend it continuously to $\overline{\Omega}$). If we replace v^* by $v^* - \Phi(g)$ and apply the argument above, we obtain the unique weak solution $\bar{u} \in H^1$ for the nonhomogeneous Dirichlet boundary condition problem:

$$\begin{cases} -\Delta u + \mu u = v^* \text{ in } \Omega \\ \qquad\quad u = g \text{ in } \partial\Omega. \end{cases}$$

Neumann Boundary Condition

Let Ω be connected and let $\partial\Omega$ be sufficiently smooth. Consider the problem with homogeneous Neumann boundary condition:

$$\begin{cases} -\Delta u + \mu u = v^* \text{ in } \Omega \\ \qquad\quad \frac{\partial u}{\partial v} = 0 \text{ in } \partial\Omega, \end{cases}$$

where v denotes the outward normal vector to $\partial\Omega$.

If $\mu > 0$, the bilinear function $B : H^1 \times H^1 \to \mathbf{R}$, defined by

$$B(u, v) = \mu \langle u, v \rangle_{L^2} + \langle \nabla u, \nabla v \rangle_{(L^2)^N},$$

is uniformly elliptic, and the arguments above allow us to prove that the problem has a unique weak solution, which is a $\bar{u} \in H^1$ such that

$$\mu \langle \bar{u}, v \rangle_{L^2} + \langle \nabla \bar{u}, \nabla v \rangle_{(L^2)^N} = \langle v^*, v \rangle_{(H^1)^*, H^1}$$

for all $v \in H^1$.

If $\mu = 0$, then B is *not* uniformly elliptic (it is not even positive definite, since $B(u, u) = 0$ when u is constant). However, by Poincaré's Inequality (part ii) of Proposition 4.14), it is possible to prove that B is uniformly elliptic when restricted to

$$V = \{ u \in H^1 : \langle u, 1 \rangle_{L^2} = 0 \},$$

which is a closed subspace of H^1. Using this fact, the problem

$$\begin{cases} -\Delta u = v^* \text{ in } \Omega \\ \frac{\partial u}{\partial v} = 0 \text{ in } \partial\Omega \end{cases}$$

has a solution (exactly one in V) whenever $v^* \in (H^1)^*$ and $\langle v^*, 1 \rangle_{(H^1)^*, H^1} = 0$.

The Obstacle Problem

Let $\phi \in H^1_0$. The set

$$C = \{ u \in H^1_0 : u(\xi) \geq \phi(\xi) \text{ for almost every } \xi \in \Omega \}$$

is a nonempty, closed and convex subset of H^1_0. With the notation introduced for the Dirichlet problem, and using Stampacchia's Theorem 4.12, we deduce that there exists a unique $\bar{u} \in C$ such that

$$\mu \langle \bar{u}, c - \bar{u} \rangle_{L^2} + \langle \nabla\bar{u}, \nabla(c - \bar{u}) \rangle_{(L^2)^N} \geq \langle v^*, c - \bar{u} \rangle_{(H^1_0)^*, H^1_0}$$

for all $c \in C$. In other words, $\bar{u}$ is a weak solution for the Dirichlet boundary value problem with an obstacle:

$$\begin{cases} -\Delta u + \mu u \geq v^* \text{ in } \Omega \\ u \geq \phi \text{ in } \Omega \\ u = 0 \text{ in } \partial\Omega. \end{cases}$$

To fix ideas, take $\mu = 0$ and $v^* = 0$, and assume u and ϕ are of class $\mathscr{C}^2$. For each nonnegative function $p \in H^1_0$ of class $\mathscr{C}^2$, the function $c = \bar{u} + p$ belongs to C, and so $\langle \nabla\bar{u}, \nabla p \rangle_{(L^2)^N} \geq 0$. Therefore, $\Delta\bar{u} \leq 0$ on Ω, and $\bar{u}$ is *superharmonic*. Moreover, if $\bar{u} > \phi$ on some open subset of Ω, then $\Delta\bar{u} \geq 0$ there. It follows that $\Delta\bar{u} = 0$, and $\bar{u}$ is *harmonic* whenever $\bar{u} > \phi$. Summarizing, u is harmonic in the *free set* (where $\bar{u} > \phi$), and both $\bar{u}$ and ϕ are superharmonic in the *contact set* (where $\bar{u} = \phi$). Under these conditions, it follows that $\bar{u}$ is the least of all superharmonic functions that lie above ϕ. For further discussion, see [69, Chap. 6].

Poisson's Equation for the p-Laplacian

Let $\Omega \subset \mathbf{R}^N$ be bounded and take $p > 1$, $\mu \in \mathbf{R}$ and $v^* \in (W_0^{1,p})^*$. Define $f : W_0^{1,p} \to \mathbf{R}$ by

$$f(u) = \frac{\mu}{p} \|u\|_{L^p}^p + \frac{1}{p} \|\nabla u\|_{(L^p)^N}^p - \langle v^*, u \rangle_{(W_0^{1,p})^*, W_0^{1,p}}.$$

Clearly, f is proper and semicontinuous. Moreover, it is strongly convex if $\mu > -1/C^2$, where C is given by Poincaré's Inequality (part i) of Proposition 4.14). By Theorem 2.19 (or, if you prefer, by Corollary 2.20), f has a unique minimizer $\bar{u}$ in $W_0^{1,p}$. Since f is differentiable, Fermat's Rule (Theorem 3.24) implies

$$\nabla f(\bar{u}) = 0.$$

With the convention that $0^{p-2} \cdot 0 = 0$ for $p > 1$, this is

$$0 = \mu \langle \|\bar{u}\|^{p-2} \bar{u}, v \rangle_{L^q, L^p} + \langle \|\nabla \bar{u}\|^{p-2} \nabla \bar{u}, \nabla v \rangle_{(L^q)^N, (L^p)^N} - \langle v^*, v \rangle_{(W_0^{1,p})^*, W_0^{1,p}}$$

for all $v \in W_0^{1,p}$. If $\mu = 0$, we recover a classical existence and uniqueness result for the *p-Laplace Equation* (see, for instance, [74, Example 2.3.1]): The function $\bar{u}$ is the unique weak solution for the problem

$$\begin{cases} -\Delta_p u = v^* \text{ in } \Omega \\ \quad\quad u = 0 \text{ in } \partial\Omega, \end{cases}$$

where the *p-Laplacian operator* $\Delta_p : W_0^{1,p} \to (W_0^{1,p})^*$ is given by

$$\Delta_p(u) = \text{div}(\|\nabla u\|^{p-2} \nabla u).$$

4.4 Sparse Solutions for Underdetermined Systems of Equations

Let A be a matrix of size $J \times N$ and let $b \in \mathbf{R}^J$. When the system $Ax = b$ is underdetermined, an important problem in signal compression and statistics (see [47, 99]) is to find its *sparsest* solutions. That is,

(P_0) $\qquad\qquad\qquad \min\{\|x\|_0 : Ax = b\},$

where $\|\cdot\|_0$ denotes the so-called *counting norm* (number of nonzero entries; not a norm). This comes from the necessity of representing rather complicated functions (such as images or other signals) in a simple and concise manner. The convex relaxation of this nonconvex problem is

(P_1) $\qquad\qquad\qquad \min\{\|x\|_1 : Ax = b\}.$

Under some conditions on the matrix A (see [48]) solutions of (P_0) can be found by solving (P_1). This can be done, for instance, by means of the *iterative shrinkage/thresholding algorithm* (ISTA), described in Example 6.26.

Related ℓ^1-regularization approaches can be found in [34, 39, 58]. The *least absolute shrinkage and selection operator* (LASSO) method [99] is closely related but slightly different: it considers a constraint of the form $\|x\|_1 \leq T$ and minimizes $\|Ax - b\|^2$. Also, it is possible to consider the problem with inequality constraints

$$\min\{\|x\|_1 : Ax \leq b\},$$

as well as the *stable signal recovery* [35], where the constraint $Ax = b$ is replaced by $\|Ax - b\| \leq \varepsilon$.

Chapter 5
Problem-Solving Strategies

Abstract Only in rare cases, can problems in function spaces be solved analytically
and exactly. In most occasions, it is necessary to apply computational methods to
approximate the solutions. In this chapter, we discuss some of the basic general
strategies that can be applied. First, we present several connections between optimization and discretization, along with their role in the problem-solving process.
Next, we introduce the idea of iterative procedures, and discuss some abstract tools
for proving their convergence. Finally, we comment some ideas that are useful to
simplify or reduce the problems, in order to make them tractable or more efficiently
solved.

5.1 Combining Optimization and Discretization

The main idea is to combine optimization techniques (based either on optimality
conditions or iterative procedures) and numerical analysis tools, which will allow us
to pass from the functional setting to a computationally tractable model and back. In
rough terms, we can identify two types of strategy, depending on the order in which
these components intervene, namely:

Discretize, Then Optimize

The strategy here consists in replacing the original problem in the infinite-
dimensional space X by a problem in a finite-dimensional subspace X_n of X^1. To
this end, the first step is to modify, reinterpret or approximate the objective function
and the constraints by a finite-dimensional model (we shall come back to this point
later). Next, we use either the optimality conditions, as those discussed in Chap. 3,

---

[1] Or even a large-dimensional X by a subspace X_n of smaller dimension.

J. Peypouquet, *Convex Optimization in Normed Spaces*,
SpringerBriefs in Optimization, DOI 10.1007/978-3-319-13710-0_5

or an iterative method, like the ones we will present in Chap. 6, for the latter. If the model is sufficiently accurate, we can expect the approximate solution x_n of the problem in X_n to be close to a real solution for the original problem in X, or, at least, that the value of the objective function in x_n be close to its minimum in X. The theoretical justification for this procedure will be discussed below.

Optimize, Then Discretize

An alternative is to use optimization techniques in order to design a method for solving the original problem directly, and then use numerical analysis techniques to implement it. Here, again, there are at least two options:

1. One option is to determine optimality conditions (see Chap. 3). These are usually translated into a functional equation (an ordinary or partial differential equation, for instance) or inequality, which is then solved numerically.
2. Another possibility is to devise an iterative procedure (this will be discussed in Chap. 6) and implement it computationally. To fix ideas, in the unconstrained case, at step n, starting from a point x_n, we identify a direction in which the objective function decreases, and advance in that direction to find the next iterate x_{n+1} such that the value of the objective function at x_{n+1} is less than the value at the previous point x_n.

5.1.1 Recovering Solutions for the Original Problem: Ritz's Method

In this section, we present a classical and very useful result that relates the infinite-dimensional problem to its finite-dimensional approximations. It is an important tool in general numerical analysis and provides a theoretical justification for the *discretize, then optimize* strategy.

Let $J : X \to \mathbf{R}$ be bounded from below. Consider a sequence (X_n) of subspaces of X such that $X_{n-1} \subset X_n$ for each $n \geq 1$, and whose union is dense in X. At the n-th iteration, given a point $x_{n-1} \in X_{n-1}$ and a tolerance $\varepsilon_n \geq 0$, find x_n such that

$$x_n \in \varepsilon_n - \operatorname{argmin}\{ J(x) : x \in X_n \} \quad \text{and} \quad J(x_n) \leq J(x_{n-1}).$$

This is always possible since J is bounded from below and $X_{n-1} \subset X_n$. Clearly, if $\varepsilon_n = 0$, the condition $J(x_n) \leq J(x_{n-1})$ holds automatically.

This procedure allows us to construct a minimizing sequence for J, along with minimizers for J. More precisely, we have the following:

Theorem 5.1 (Ritz's Theorem). *If J is upper-semicontinuous and $\lim\limits_{n \to \infty} \varepsilon_n = 0$, then $\lim\limits_{n \to \infty} J(x_n) = \inf\limits_{x \in X} J(x)$. If, moreover, J is continuous, then every limit point of (x_n) minimizes J.*

Proof. Fix $x \in X$ and $\delta > 0$. There exists $y_0 \in X$ such that

$$J(y_0) \leq J(x) + \delta/3.$$

Since J is upper-semicontinuous at y_0, there is $\rho > 0$ such that

$$J(y) < J(y_0) + \delta/3$$

whenever $\|y - y_0\| < \rho$. Using the density hypothesis and the fact that $\lim\limits_{n \to \infty} \varepsilon_n = 0$, we can find $N \in \mathbf{N}$ and $z_N \in X_N$ such that $\|z_N - y_0\| < \rho$, and $\varepsilon_N \leq \delta/3$. We deduce that

$$J(x_n) \leq J(x_N) \leq J(z_N) + \varepsilon_N < J(y_0) + 2\delta/3 \leq J(x) + \delta$$

for every $n \geq N$. Since this holds for each $x \in X$, we obtain the minimizing property. The last part follows from Proposition 2.8. $\qquad \square$

Typically, the finite-dimensional space X_n is spanned by a finite (not necessarily n) number of elements of X, that are chosen considering the nature of the problem and its solutions.

5.1.2 Building the Finite-Dimensional Approximations

Most optimization problems in infinite-dimensional spaces (see Chap. 4 for some examples) share the following properties:

1. The *variables* are functions whose domain is some subset of $\mathbf{R}^N$.
2. The objective function involves an integral functional depending on the variables and possibly their derivatives.
3. The constraints are given by functional equations or inequalities.

The most commonly used discretization schemes for these kinds of problems, use a finite number of elements selected from a Schauder basis for the space and consider approximations in the subset spanned by these elements. Here are a few examples:

Finite Differences: Draw a grid in the domain, partitioning it into small N-dimensional blocks.

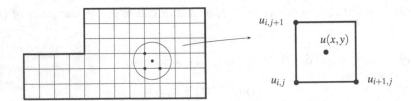

Represent the functions on each block by their values on a prescribed node, producing piecewise constant approximations. Integrals are naturally approximated by a Riemann sum. Partial derivatives are replaced by the corresponding difference quotients. These can be, for instance, *one-sided* $\frac{\partial u}{\partial x}(x,y) \sim \frac{u_{i+1,j}-u_{i,j}}{h}$; or *centered* $\frac{\partial u}{\partial x}(x,y) \sim \frac{u_{i+1,j}-u_{i-1,j}}{2h}$. The dimension of the approximating space is the number of nodes and the new variables are vectors whose components are the values of the original variable (a function) on the nodes.

Finite Elements: As before, the domain is partitioned into small polyhedral units, whose geometry may depend on the dimension and the type of approximation. On each unit, the function is most often represented by a polynomial. Integrals and derivatives are computed accordingly. The dimension of the approximating space depends on the number of subdomains in the partition and the type of approximation (for instance, the degree of the polynomials). For example:

1. If $N = 1$, the Trapezoidal Rule for computing integrals is a approximation by polynomials of order 1, while Simpson's Rule uses polynomials of order 2.
2. When $N = 2$, it is common to partition the domain into triangles. If function values are assigned to the vertices, it is possible to construct an affine interpolation. This generates the graph of a continuous function whose derivatives, which are constant on each triangle, and integral are easily computed.

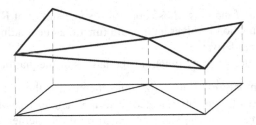

The method of finite elements has several advantages: First, reasonably smooth open domains can be well approximated by triangulation. Second, partitions can be very easily refined, and one can choose not to do this refinement uniformly throughout the domain, but only in specific areas where more accuracy is required. Third, since the elements have very small support, the method is able to capture local behavior in a very effective manner. In particular, if the support of the solution is localized, the finite element scheme will tend to identify this fact very quickly. Overall, this method offers a good balance between simplicity, versatility, and precision. One possible drawback is that it usually requires very fine partitions in order to represent smooth functions. The interested reader is referred to [28, 41], or [90].

Fourier Series: Trigonometric polynomials provide infinitely differentiable approximations to piecewise smooth functions. One remarkable property is that their integrals and derivatives can be explicitly computed and admit very accurate approximations. Another advantage is that functions in the basis can all be generated from any *one* of them, simply by a combination of translations and contractions, which is a very positive feature when it comes to implementation. To fix the ideas and see the difference with the previous methods, suppose the domain is a bounded interval, interpreted as a time horizon. The Fourier coefficients are computed by means of projections onto sinusoidal functions with different frequencies. The dimension of the approximating space is the number of Fourier coefficients computed. In some sense, the original function is replaced by a representation that takes into account its main oscillation frequencies. There are two main drawbacks in taking the standard Fourier basis. First comes its lack of localization. A very sparse function (a function with a very small support) may have a large number of nonnegligible Fourier coefficients. The second weak point is that the smooth character of the trigonometric functions offers a poor approximation around discontinuities and produces some local instabilities such as the Gibbs Phenomenon. More details on Fourier analysis and series can be found in [56, 100], or still [18].

Wavelets: This term gathers a broad family of methods involving representation of functions with respect to certain bases. The common and distinctive features are the following:

- *Localization:* Each element in the basis may be chosen either with arbitrarily small support, or having very small values outside an arbitrarily small domain. This makes them ideal for identifying peaks and other types of local behavior. Just like the finite elements, they reveal sparsity very quickly.
- *Simple representation:* All the functions in the basis are constructed from *one* particular element, called the *mother wavelet*, by translations and contractions.
- *Customized smoothness:* One can choose pretty much any degree of smoothness ranging from piecewise constant (such as the *Haar* basis) to infinitely differentiable (like the so-called *Mexican hat*) functions, depending on the qualitative nature of the functions to approximate.
- *Capture oscillating behavior:* The shape of the elements of the basis makes it possible to identify the main oscillating frequencies of the functions.

Wavelet bases share several good properties of finite elements and Fourier series. For further reference, the reader may consult [46] or [18].

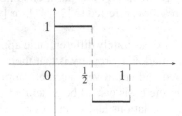

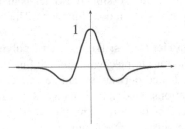

The Haar mother wavelet   The *Mexican hat* mother wavelet

5.2 Iterative Procedures

An *iterative algorithm* on X is a procedure by which, starting from a point $z_0 \in X$, and using a family (F_n) of functions from X to X, we produce a sequence (z_n) of points in X by applying the rule

$$z_{n+1} = F_n(z_n)$$

for each $n \geq 0$.

In the search for minimizers of a function f, a useful idea is to follow a certain direction in order to find a point where f has a smaller value. In the process, we construct a sequence that is meant to minimize f. We shall come back to this point in Chap. 6.

A Useful Criterion for Weak Convergence in Hilbert Space

It is possible to establish the strong or weak convergence of a sequence by using the definition, provided one knows the limit beforehand. Otherwise, one must rely on different tools. For example, the Cauchy property provides a convergence criterion in Banach spaces. The following result is useful to prove the weak convergence of a sequence in a Hilbert space without *a priori* knowledge of the limit.

Lemma 5.2 (Opial's Lemma). *Let S be a nonempty subset of a Hilbert space H and let (z_n) be a sequence in H. Assume*

i) *For each $u \in S$ there exists $\lim_{n \to \infty} \|z_n - u\|$; and*

ii) *Every weak limit point of (z_n) belongs to S.*

Then (z_n) converges weakly as $n \to \infty$ to some $\bar{u} \in S$.

Proof. Since (z_n) is bounded, it suffices to prove that it has at most one weak limit point, by Corollary 1.25. Let $z_{k_n} \rightharpoonup \hat{z}$ and $z_{m_n} \rightharpoonup \check{z}$. Since $\hat{z}, \check{z} \in S$, the limits $\lim_{n \to \infty} \|z_n - \hat{z}\|$ and $\lim_{n \to \infty} \|z_n - \check{z}\|$ exist. Since

$$\|z_n - \hat{z}\|^2 = \|z_n - \check{z}\|^2 + \|\check{z} - \hat{z}\|^2 + 2\langle z_n - \check{z}, \check{z} - \hat{z} \rangle,$$

by passing to appropriate subsequences we deduce that $\hat{z} = \check{z}$. ☐

The formulation and proof presented above for Lemma 5.2 first appeared in [19]. Part of the ideas are present in [84], which is usually given as the standard reference.

Qualitative Effect of Small Computational Errors

The minimizing property of sequences of approximate solutions is given by Ritz's Theorem 5.1. However, when applying an iterative algorithm, numerical errors typically appear in the computation of the iterates. It turns out that small computational errors do not alter the qualitative behavior of sequences generated by nonexpansive algorithms. Most convex optimization algorithms are nonexpansive. We have the following:

Lemma 5.3. *Let (F_n) be a family of nonexpansive functions on a Banach space X and assume that every sequence (z_n) satisfying $z_n = F_n(z_{n-1})$ converges weakly (resp. strongly). Then so does every sequence $(\widehat{z}_n)$ satisfying*

$$\widehat{z}_n = F_n(\widehat{z}_{n-1}) + \varepsilon_n \tag{5.1}$$

provided $\sum_{n=1}^{\infty} \|\varepsilon_n\| < +\infty$.

Proof. Let τ denote either the strong or the weak topology. Given $n > k \geq 0$, define

$$\Pi_k^n = F_n \circ \cdots \circ F_{k+1}.$$

In particular, $\tau - \lim_{n \to \infty} \Pi_k^n(z)$ exists for each $k \in \mathbf{N}$ and $z \in X$. Let $(\widehat{z}_n)$ satisfy (5.1) and set

$$\zeta_k = \tau - \lim_{n \to \infty} \Pi_k^n(\widehat{z}_k).$$

Since

$$\|\Pi_{k+h}^n(\widehat{z}_{k+h}) - \Pi_k^n(\widehat{z}_k)\| = \|\Pi_{k+h}^n(\widehat{z}_{k+h}) - \Pi_{k+h}^n \circ \Pi_k^{k+h}(\widehat{z}_k)\|$$
$$\leq \|\widehat{z}_{k+h} - \Pi_k^{k+h}(\widehat{z}_k)\|,$$

we have

$$\|\zeta_{k+h} - \zeta_k\| \leq \lim_{n \to \infty} \|\Pi_{k+h}^n(\widehat{z}_{k+h}) - \Pi_k^n(\widehat{z}_k)\| \leq \|\widehat{z}_{k+h} - \Pi_k^{k+h}(\widehat{z}_k)\|.$$

But, by definition,

$$\|\widehat{z}_{k+h} - \Pi_k^{k+h}(\widehat{z}_k)\| \le \sum_{j=k}^{\infty} \|\varepsilon_j\|,$$

and so, it tends to zero uniformly in h as $k \to \infty$. Therefore (ζ_k) is a Cauchy sequence that converges strongly to a limit ζ. Write

$$\widehat{z}_{k+h} - \zeta = [\widehat{z}_{k+h} - \Pi_k^{k+h}(\widehat{z}_k)] + [\Pi_k^{k+h}(\widehat{z}_k) - \zeta_k] + [\zeta_k - \zeta]. \qquad (5.2)$$

Given $\varepsilon > 0$ we can take k large enough so that the first and third terms on the right-hand side of (5.2) are less than ε in norm, uniformly in h. Next, for such k, we let $h \to \infty$ so that the second term converges to zero for the topology τ. $\square$

Further robustness properties for general iterative algorithms can be found in [5, 6] and [7].

5.3 Problem Simplification

There are several ways to simplify optimization problems. Here, we mention two important techniques: one consists in passing from the constrained setting to an unconstrained one; the other allows us to deal with different sources of difficulty separately.

5.3.1 Elimination of Constraints

Constrained optimization problems are, in general, hard to solve. In some occasions, it is possible to reformulate the problem in an equivalent (or nearly equivalent), unconstrained version. We shall present two different strategies, which are commonly used. Of course, this list is not exhaustive (see, for instance, [95]).

Penalization

Suppose, for simplicity, that $f : X \to \mathbf{R}$ is convex and continuous, and consider the constrained problem

$$(P) \qquad \min\{f(x) : x \in C\} = \min\{f(x) + \delta_C(x) : x \in X\},$$

where $C \subset X$ is nonempty, closed and convex. If one replaces δ_C by a function that is similar in some sense, it is natural to expect the resulting problem to have similar solutions as well.

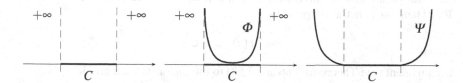

A (not necessarily positive) function looking like Φ in the figure above is usually referred to as an *interior penalization* function for C. It satisfies $\text{int}(C) \subset \text{dom}(\Phi)$, $\overline{\text{dom}(\Phi)} = C$, and $\lim_{y \to x} \|\nabla \Phi(y)\| = +\infty$ for all x on the boundary of C, denoted by ∂C. A *Legendre* function is an interior penalization function that is differentiable and strictly convex in $\text{int}(C)$ (see [91, Chap. 26]). A *barrier* is a Legendre function such that $\lim_{y \to x} \Phi(y) = +\infty$ for all $x \in \partial C$. In turn, a convex, nonnegative, everywhere defined function Ψ such that $\Psi^{-1}(0) = C$ is an *exterior penalization* function, since it act only outside of C.

Under suitable conditions, solutions for the (rather smoothly constrained) problem

$$(P_\varepsilon) \qquad\qquad \min\{ f(x) + \varepsilon \Phi(x) : x \in X \},$$

for $\varepsilon > 0$, exist and lie in $\text{int}(C)$. One would reasonably expect them to approximate solutions of (P), as $\varepsilon \to 0$. Indeed, by weak lower-semicontinuity, it is easy to see that, if x_ε minimizes $f + \varepsilon \Phi$ on X, then every weak limit point of (x_ε), as $\varepsilon \to 0$, minimizes f on C. Similarly, if x_β is a solution for the unconstrained problem

$$(P_\beta) \qquad\qquad \min\{ f(x) + \beta \Psi(x) : x \in X \},$$

with $\beta > 0$, then every weak limit point of (x_β), as $\beta \to +\infty$, belongs to C and minimizes f on C.

Remark 5.4. Loosely speaking, we replace the function $f + \delta_C$ by a family (f_ν) of functions in such a way that f_ν is reasonably similar to $f + \delta_C$ if the parameter ν is close to some value ν_0.

Multiplier Methods

As in Sect. 3.7.2, consider proper, lower-semicontinuous and convex functions $f, g_1, \ldots, g_m : X \to \mathbf{R} \cup \{+\infty\}$, and focus on the problem of minimizing a convex function $f : X \to \mathbf{R} \cup \{+\infty\}$ over the set C defined by

$$C = \{ x \in X : g_i(x) \le 0 \text{ for } i = 1, \ldots, m \}.$$

Suppose we are able to solve the dual problem and find a Lagrange multiplier $\hat{\lambda} \in \mathbf{R}^m_+$. Then, the constrained problem

$$\min\{ f(x) : x \in C \}$$

is equivalent (see Theorems 3.66 and 3.68) to the unconstrained problem

$$\min \left\{ f(x) + \sum_{i=1}^{m} \hat{\lambda}_i g_i(x) : x \in X \right\}.$$

One can proceed either:

- Subsequently: solving the (finite-dimensional) dual problem first, and then using the Lagrange multiplier to state and solve an unconstrained primal problem; or

- Simultaneously: approach the primal-dual solution in the product space. This, in turn, can be performed in two ways:

  - Use optimality conditions for the saddle-point problem. The primal-dual solution must be a critical point of the Lagrangian, if it is differentiable;
  - Devise an algorithmic procedure based on iterative minimization with respect to the primal variable x and maximization on the dual variable λ. We shall comment this method in Chap. 6.

Similar approaches can be used for the affinely constrained problem where

$$C = \{x \in X : Ax = b\}$$

(see Theorem 3.63). For instance, the constrained problem is equivalent to an unconstrained problem in view of Remark 3.70, and the optimality condition is a system of inclusions: $A\hat{x} = b$, and $-A^*y^* \in \partial f(\hat{x})$ for $(\hat{x}, y^*) \in X \times Y^*$.

5.3.2 Splitting

In many practical cases, one can identify *components* either in the objective function, the variable space, or the constraint set. In such situations, it may be useful to focus on the different components independently, and then use the partial information to find a point which is closer to being a solution for the original problem.

Additive and Composite Structure

In some occasions, the objective function f is structured as a sum of functions $f_1, \ldots, f_K$, each of which has special properties. One common strategy consists in performing minimization steps on each f_k $(k = 1, \ldots, K)$ independently (this can be done either simultaneously or successively), and then using some averaging or projection procedure to consolidate the information. Let us see some model situations, some of which we will revisit in Chap. 6:

Separate variables in a product space: The space X is the product of two spaces X_1 and X_2, where $f_i : X_i \to \mathbf{R} \cup \{+\infty\}$ for $i = 1, 2$, and the corresponding variables x_1 and x_2 are linked by a simple relation, such as an affine equation.

Example 5.5. For the linear-quadratic optimal control problem described in Sect. 4.2.3, we can consider the variables $u \in L^p(0, T; \mathbf{R}^M)$ and $y \in \mathscr{C}([0, T]; \mathbf{R}^N)$, related by the Variation of Parameters Formula (4.1). □

When independent minimization is simple: The functions f_1 and f_2 depend on the same variable but minimizing the sum is difficult, while minimizing each function separately is relatively simple.

Example 5.6. Suppose one wishes to find a point in the intersection of two closed convex subsets, C_1 and C_2, of a Hilbert space. This is equivalent to minimizing $f = \delta_{C_1} + \delta_{C_2}$. We can minimize δ_{C_i} by projecting onto C_i. It may be much simpler to project onto C_1 and C_2 independently, than to project onto $C_1 \cap C_2$. This method of *alternating projections* has been studied in [40, 61, 102] among others. □

Different methods for different regularity: For instance, suppose that $f_1 : X \to \mathbf{R}$ is smooth, while $f_2 : X \to \mathbf{R} \cup \{+\infty\}$ is proper, convex and lower-semicontinuous. We shall see, in Chap. 6, that some methods are well adapted to nonsmooth functions, whereas other methods require certain levels of smoothness to be effective.

Example 5.7. Suppose we wish to minimize an everywhere defined, smooth function f over a closed, convex set C. In Hilbert spaces, the projected gradient algorithm (introduced by [59] and [73]) consists in performing an iteration of the gradient method (see Sect. 6.3) with respect to f, followed by a projection onto C. □

Composite structure: If $A : X \to Y$ is affine and $f : Y \to \mathbf{R} \cup \{+\infty\}$ is convex, then $f \circ A : X \to \mathbf{R} \cup \{+\infty\}$ is convex (see Example 2.13). In some cases, and depending on f and A, it may be useful to consider a constrained problem in the product space. More precisely, the problems

$$\min_{x \in X} \{ f(Ax) \} \qquad \text{and} \qquad \min_{(x,y) \in X \times Y} \{ f(y) : Ax = y \}$$

are equivalent. The latter has an additive structure and the techniques presented above can be used, possibly in combination with iterative methods. This is particularly useful when f has a simple structure, but $f \circ A$ is complicated.

Example 5.8. As seen in Sect. 4.3.3, the optimality condition for minimizing the square of the L^2 norm of the gradient is a second-order partial differential equation. □

We shall come back to these techniques in Chap. 6.

Chapter 6
Keynote Iterative Methods

Abstract In this chapter we give an introduction to the basic (sub)gradient-based methods for minimizing a convex function on a Hilbert space. We pay special attention to the proximal point algorithm and the gradient method, which are interpreted as time discretizations of the steepest descent differential inclusion. Moreover, these methods, along with some extensions and variants, are the building blocks for other — more sophisticated — methods that exploit particular features of the problems, such as the structure of the feasible (or constraint) set. The choice of proximal- or gradient-type schemes depends strongly on the regularity of the objective function.

Throughout this chapter, H is a real Hilbert space.

6.1 Steepest Descent Trajectories

Let $f : H \to \mathbf{R}$ be a continuously differentiable function with Lipschitz-continuous gradient. By the Cauchy-Lipschitz-Picard Theorem, for each $x_0 \in H$, the ordinary differential equation

$$(ODE) \qquad \begin{cases} x(0) = x_0 \\ -\dot{x}(t) = \nabla f(x(t)), \quad t > 0, \end{cases}$$

has a unique solution. More precisely, there is a unique continuously differentiable function $x : [0, +\infty) \to H$ such that $x(0) = x_0$ and $-\dot{x}(t) = \nabla f(x(t))$ for all $t > 0$. To fix the ideas, let us depict this situation in $\mathbf{R}^2$:

© The Author(s) 2015
J. Peypouquet, *Convex Optimization in Normed Spaces*,
SpringerBriefs in Optimization, DOI 10.1007/978-3-319-13710-0_6

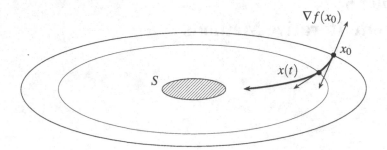

At each instant t, the velocity $\dot{x}(t)$ points towards the interior of the sublevel set $\Gamma_{f(x(t))}(f)$. Observe also that the stationary points of (ODE) are exactly the critical points of f (the zeroes of ∇f). Moreover, the function f decreases along the solutions, and does so strictly unless it bumps into a critical point, since

$$\frac{d}{dt}f(x(t)) = \langle \nabla f(x(t)), \dot{x}(t) \rangle = -\|\dot{x}(t)\|^2 = -\|\nabla f(x(t))\|^2.$$

If f is convex, one would reasonably expect the trajectories to minimize f as $t \to \infty$. Indeed, by convexity, we have (see Proposition 3.10)

$$f(z) - f(x(t)) \geq \langle \nabla f(x(t)), z - x(t) \rangle = \langle \dot{x}(t), x(t) - z \rangle = \frac{d}{dt}h_z(t),$$

where $h_z(t) = \frac{1}{2}\|x(t) - z\|^2$. Integrating on $[0, T]$ and recalling that $t \mapsto f(x(t))$ is nonincreasing, we deduce that

$$f(x(T)) \leq f(z) + \frac{\|x_0 - z\|^2}{2T},$$

and we conclude that

$$\lim_{t \to \infty} f(x(t)) = \inf(f).$$

If, moreover, $S = \operatorname{argmin}(f) \neq \emptyset$ and we take $\bar{z} \in S$, then

$$\frac{d}{dt}h_{\bar{z}}(t) = \langle \dot{x}(t), x(t) - \bar{z} \rangle = -\langle \nabla f(x(t)) - \nabla f(\bar{z}), x(t) - \bar{z} \rangle \leq 0$$

and $\lim_{t \to \infty} \|x(t) - \bar{z}\|$ exists. Since f is continuous and convex, it is weakly lower-semicontinuous, by Proposition 2.17. Therefore, every weak limit point of $x(t)$ as $t \to \infty$ must minimize f, by Proposition 2.8. We conclude, by Opial's Lemma 5.2, that $x(t)$ converges weakly to some $\bar{z} \in S$ as $t \to \infty$.

It is possible to prove that if $f : H \to \mathbf{R} \cup \{+\infty\}$ is proper, lower-semicontinuous and convex, then the *Steepest Descent Differential Inclusion*

$$(DI) \qquad \begin{cases} x(0) = x_0 \\ -\dot{x}(t) \in \partial f(x(t)), \quad t > 0, \end{cases}$$

has similar properties: for each $x_0 \in \overline{\mathrm{dom}(f)}$, (DI) has a unique absolutely continuous solution $x : [0, +\infty) \to H$, that it satisfies $\lim_{t \to \infty} f(x(t)) = \inf(f)$. Further, if $S \neq \emptyset$, then $x(t)$ converges weakly to a point in S. The proof (especially the existence) is more technical and will be omitted. The interested reader may consult [87] for further discussion, or also [29], which is the original source.

If we discretize (ODE) by the method of finite differences, we first take a sequence (λ_n) of positive parameters called the *step sizes*, set $\sigma_n = \sum_{k=1}^{n} \lambda_k$, and partition the interval $[0, +\infty)$ as:

$$\begin{array}{ccccccc} \lambda_1 & \lambda_2 & \lambda_3 & & & \lambda_n & \\ \vdash\!\!-\!\!\!+\!\!-\!\!+\!\!-\!\!-\!\!+\!\!-\!\!-\!\!-\!\!-\!\!-\!\!-\!\!-\!\!-\!\!+\!\!-\!\!+\!\!-\!\!\to & & & & & & \\ 0 & \sigma_1 \ \sigma_2 & \sigma_3 & \cdots & \sigma_{n-1} & \sigma_n & \end{array}$$

In order to recover the asymptotic properties of (ODE) as $t \to \infty$, we suppose that $\sigma_n \to \infty$ as $n \to \infty$, which is equivalent to $(\lambda_n) \in \ell^1$.

Then $\dot{x}(t)$ is approximated by

$$\frac{x_n - x_{n-1}}{\lambda_n},$$

where the sequence (x_n) must be determined. On the other hand, it is natural to approximate the term $\nabla f(x(t))$ either by $\nabla f(x_{n-1})$, or by $\nabla f(x_n)$, which correspond to evaluating the gradient at the time corresponding to the left, or the right, end of the interval $[\sigma_{n-1}, \sigma_n]$, respectively. The first option produces an explicit update rule:

$$-\frac{x_n - x_{n-1}}{\lambda_n} = \nabla f(x_{n-1}) \qquad \Longleftrightarrow \qquad x_n = x_{n-1} - \lambda_n \nabla f(x_{n-1}),$$

known as the *gradient method*, originally devised by Cauchy [36]. This method finds points in S if f is sufficiently regular and the step sizes are properly selected. The second option, namely $\nabla f(x(t)) \sim \nabla f(x_n)$, gives an implicit update rule:

$$-\frac{x_n - x_{n-1}}{\lambda_n} = \nabla f(x_n) \qquad \Longleftrightarrow \qquad x_n + \lambda_n \nabla f(x_n) = x_{n-1},$$

that is known as the *proximal point algorithm*, and was introduced by Martinet [76] much more recently. Despite the difficulty inherent to the implementation of the implicit rule, this method has remarkable stability properties, and can be applied to non-differentiable (even discontinuous) functions successfully. These two algorithms are the building blocks for a wide spectrum of iterative methods in optimization. The remainder of this chapter is devoted to their study.

6.2 The Proximal Point Algorithm

In what follows, $f : H \to \mathbf{R} \cup \{+\infty\}$ is a proper, lower-semicontinuous convex function. We denote the optimal value

$$\alpha = \inf\{f(u) : u \in H\},$$

which, in principle, may be $-\infty$; and the solution set

$$S = \operatorname{argmin}(f),$$

which, of course, may be empty. To define the proximal point algorithm, we consider a sequence (λ_n) of positive numbers called the *step sizes*.

Proximal Point Algorithm: Begin with $x_0 \in H$. For each $n \geq 0$, given a step size λ_n and the state x_n, define the state x_{n+1} as the unique minimizer of the Moreau–Yosida Regularization $f_{(\lambda_n, x_n)}$ of f, which is the proper, lower-semicontinuous and strongly convex function defined as

$$f_{(\lambda_n, x_n)}(z) = f(z) + \frac{1}{2\lambda_n}\|z - x_n\|^2$$

(see Sect 3.5.4). In other words,

$$\{x_{n+1}\} = \operatorname{argmin}\left\{ f(z) + \frac{1}{2\lambda_n}\|z - x_n\|^2 : z \in H \right\}. \qquad (6.1)$$

Moreover, according to the Moreau–Rockafellar Theorem,

$$0 \in \partial f_{(\lambda_n, x_n)}(x_{n+1}) = \partial f(x_{n+1}) + \frac{x_{n+1} - x_n}{\lambda_n},$$

or, equivalently,

$$-\frac{x_{n+1} - x_n}{\lambda_n} \in \partial f(x_{n+1}). \qquad (6.2)$$

Using this property, one can interpret the *proximal iteration* $x_n \mapsto x_{n+1}$ as an implicit discretization of the differential inclusion

$$-\dot{x}(t) \in \partial f(x(t)) \qquad t > 0.$$

Inclusion (6.2) can be written in *resolvent* notation (see (3.14)) as:

$$x_{n+1} = (I + \lambda_n \partial f)^{-1}(x_n).$$

A sequence (x_n) generated following the proximal point algorithm is a *proximal sequence*. The stationary points of a proximal sequence are the minimizers of the objective function f, since, clearly,

$$x_{n+1} = x_n \qquad \text{if, and only if,} \qquad 0 \in \partial f(x_{n+1}).$$

6.2.1 Basic Properties of Proximal Sequences

We now study the basic properties of proximal sequences, which hold for any (proper, lower-semicontinuous and convex) f. As we shall see, proximal sequences minimize f. Moreover, they converge weakly to a point in the solution set S, provided it is nonempty.

In the first place, the very definition of the algorithm (see (6.1)) implies

$$f(x_{n+1}) + \frac{1}{2\lambda_n}\|x_{n+1} - x_n\|^2 \leq f(x_n)$$

for each $n \geq 0$. Therefore, the sequence $(f(x_n))$ is nonincreasing. In fact, it decreases *strictly* as long as $x_{n+1} \neq x_n$. In other words, at each iteration, the point x_{n+1} will be different from x_n only if a sufficient reduction in the value of the objective function can be attained.

Inclusion (6.2) and the definition of the subdifferential together give

$$f(u) \geq f(x_{n+1}) + \left\langle -\frac{x_{n+1} - x_n}{\lambda_n}, u - x_{n+1} \right\rangle.$$

for each $u \in H$. This implies

$$2\lambda_n [f(x_{n+1}) - f(u)] \leq 2\langle x_{n+1} - x_n, u - x_{n+1}\rangle$$
$$= \|x_n - u\|^2 - \|x_{n+1} - u\|^2 - \|x_{n+1} - x_n\|^2$$
$$\leq \|x_n - u\|^2 - \|x_{n+1} - u\|^2. \tag{6.3}$$

Summing up for $n = 0, \ldots, N$ we deduce that

$$2\sum_{n=0}^{N} \lambda_n [f(x_{n+1}) - f(u)] \leq \|x_0 - u\|^2 - \|x_{N+1} - u\|^2 \leq \|x_0 - u\|^2$$

for each $N \in \mathbf{N}$. Since the sequence $(f(x_n))$ is nonincreasing, we deduce that

$$2\sigma_N(f(x_{N+1}) - f(u)) \leq \|x_0 - u\|^2,$$

where we have written

$$\sigma_n = \sum_{k=0}^{n} \lambda_k.$$

Gathering all this information we obtain the following:

Proposition 6.1. *Let (x_n) be a proximal sequence.*

i) *The sequence $(f(x_n))$ is nonincreasing. Moreover, it decreases strictly as long as $x_{n+1} \neq x_n$.*

ii) *For each $u \in H$ and $n \in \mathbf{N}$, we have*

$$2\sigma_n(f(x_{n+1}) - f(u)) \leq \|x_0 - u\|^2.$$

In particular,

$$f(x_{n+1}) - \alpha \leq \frac{d(x_0, S)^2}{2\sigma_n}.$$

iii)*If $(\lambda_n) \notin \ell^1$, then $\lim_{n \to \infty} f(x_n) = \alpha$.*

As mentioned above, we shall see (Theorem 6.3) that proximal sequences always converge weakly. However, if a proximal sequence (x_n) happens to converge strongly (see Sect. 6.2.2), it is possible to obtain a convergence rate of $\lim_{n \to \infty} \sigma_n(f(x_{n+1}) - \alpha) = 0$, which is faster than the one predicted in part ii) of the preceding proposition. This was proved in [60] (see also [85, Sect. 2] for a simplified proof).

Concerning the sequence (x_n) itself, we have the following:

Proposition 6.2. *Let (x_n) be a proximal sequence with $(\lambda_n) \notin \ell^1$. Then:*

i) *Every weak limit point of the sequence (x_n) must lie in S.*

ii) *If $x^* \in S$, then the sequence $(\|x_n - x^*\|)$ is nonincreasing. As a consequence, $\lim_{n \to \infty} \|x_n - x^*\|$ exists.*

In particular, the sequence (x_n) is bounded if, and only if, $S \neq \emptyset$.

Proof. For part i), if (x_{k_n}) converges weakly to $\bar{x}$, then

$$f(\bar{x}) \leq \liminf_{n \to \infty} f(x_{k_n}) = \lim_{n \to \infty} f(x_n) \leq \inf\{f(u) : u \in H\},$$

by the weak lower-semicontinuity of f. This implies $\bar{x} \in S$. Part ii) is obtained by replacing $u = x^*$ in (6.3) since $f(x_n) \geq f(x^*)$ for all $n \in \mathbf{N}$. □

Proposition 6.2, along with Opial's Lemma 5.2, allow us to establish the weak convergence of proximal sequences:

Theorem 6.3. *Let (x_n) be a proximal sequence generated using a sequence of step sizes $(\lambda_n) \notin \ell^1$. If $S \neq \emptyset$, then (x_n) converges weakly, as $n \to \infty$, to a point in S. Otherwise, $\lim_{n \to \infty} \|x_n\| = \infty$.*

Consistency of the Directions

As seen above, proximal sequences converge weakly under very general assumptions. The fact that the computation of the point x_{k+1} involves information on the point x_{n+1} itself, actually implies that, in some sense, the future history of the sequence will be taken into account as well.

For this reason, the evolution of proximal sequences turns out to be very stable. More precisely, the displacement $x_{n+1} - x_n$ always forms an acute angle both with the previous displacement $x_n - x_{n-1}$ and with any vector pointing towards the minimizing set S.

Proposition 6.4. *Let (x_n) be a proximal sequence. If $x_{n+1} \neq x_n$, then*

$$\langle x^{n+1} - x^n, x^n - x^{n-1} \rangle > 0.$$

Moreover, if $x_{n+1} \neq x_n$ and $\hat{x} \in S$ then

$$\langle x^{n+1} - x^n, \hat{x} - x^n \rangle > 0.$$

Proof. According to (6.2), we have

$$-\frac{x_{n+1} - x_n}{\lambda_n} \in \partial f(x_{n+1}) \qquad \text{and} \qquad -\frac{x_n - x_{n-1}}{\lambda_{n-1}} \in \partial f(x_n).$$

By monotonicity,

$$\left\langle \frac{x_{n+1} - x_n}{\lambda_n} - \frac{x_n - x_{n-1}}{\lambda_{n-1}}, x_{n+1} - x_n \right\rangle \leq 0.$$

In other words,

$$\langle x_{n+1} - x_n, x_n - x_{n-1} \rangle \geq \frac{\lambda_{n-1}}{\lambda_n} \|x_{n+1} - x_n\|^2.$$

In particular, $\langle x_{n+1} - x_n, x_n - x_{n-1} \rangle > 0$ whenever $x_{n+1} \neq x_n$. For the second inequality, if $\hat{x} \in S$, then $0 \in \partial f(\hat{x})$. Therefore

$$\langle x_{n+1} - x_n, \hat{x} - x_n \rangle = \langle x_{n+1} - x_n, \hat{x} - x_{n+1} \rangle + \|x_{n+1} - x_n\|^2 > 0,$$

as long as $x_{n+1} \neq x_n$, by monotonicity. $\qquad \square$

In conclusion, proximal sequences cannot go "back and forth;" they always move in the direction of S.

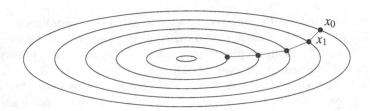

6.2.2 *Strong Convergence and Finite-Time Termination*

Theorem 6.3 guarantees the weak convergence of proximal sequences. A natural question—already posed by R.-T. Rockafellar in [93]—is whether or not the convergence is always strong. A negative answer was provided by O. Güler in [60] based on an example given by J.-B. Baillon in [20] for the differential inclusion $-\dot{x}(t) \in \partial f(x(t)), t > 0$. However, under additional assumptions, the proximal sequences can be proved to converge strongly. We shall present some examples here. Some variations of the proximal algorithm also converge strongly (see, for instance, [98]).

Strong Convexity and Expansive Subdifferential

Recall from Proposition 3.23 that if $f : H \to \mathbf{R} \cup \{+\infty\}$ is strongly convex with parameter $r > 0$, then ∂f is strongly monotone with parameter r, which means that

$$\langle x^* - y^*, x - y \rangle \geq r\|x - y\|^2$$

whenever $x^* \in \partial f(x)$ and $y^* \in \partial f(y)$. Observe also that the latter implies that ∂f is *r-expansive*, in the sense that

$$\|x^* - y^*\| \geq r\|x - y\|$$

whenever $x^* \in \partial f(x)$ and $y^* \in \partial f(y)$.

When applied to strongly convex functions, the proximal point algorithm generates strongly convergent sequences.

Proposition 6.5. *Let (x_n) be a proximal sequence with $(\lambda_n) \notin \ell^1$. If f is strongly convex with parameter r, then S is a singleton $\{\hat{x}\}$ and*

$$\|x_{n+1} - \hat{x}\| \leq \|x_0 - \hat{x}\| \prod_{k=0}^{n} (1 + r\lambda_k)^{-1}. \tag{6.4}$$

In particular, (x_n) converges strongly, as $n \to \infty$, to the unique $\hat{x} \in S$.

Proof. Since $0 \in \partial f(\hat{x})$ and $-\frac{x_{n+1}-x_n}{\lambda_n} \in \partial f(x_{n+1})$, the strong convexity implies

$$\langle x_{n+1} - x_n, \hat{x} - x_{n+1} \rangle \geq r\lambda_n \|\hat{x} - x_{n+1}\|^2$$

for all n. The equality $2\langle \zeta, \xi \rangle = \|\zeta + \xi\|^2 - \|\zeta\|^2 - \|\xi\|^2$ gives

$$\|\hat{x} - x_n\|^2 - \|x_{n+1} - x_n\|^2 \geq (1 + 2r\lambda_n)\|\hat{x} - x_{n+1}\|^2.$$

But $\|x_{n+1} - x_n\| \geq r\|\hat{x} - x_{n+1}\|$ by the expansive property of the subdifferential of a strongly convex function. We deduce that

$$\|\hat{x} - x_n\|^2 \geq (1 + 2r\lambda_n + r^2\lambda_n^2)\|\hat{x} - x_{n+1}\|^2$$

and so

$$\|\hat{x} - x_{n+1}\| \leq (1 + r\lambda_n)^{-1}\|\hat{x} - x_n\|.$$

Proceeding inductively we obtain (6.4). To conclude, we use the fact that $\prod_{k=0}^{\infty}(1 + r\lambda_k)$ tends to ∞ if, and only if, $(\lambda_n) \notin \ell^1$. $\qquad\square$

It is possible to weaken the strong convexity assumption provided the step sizes are sufficiently large. More precisely, we have the following:

Proposition 6.6. *Let (x_n) be a proximal sequence with $(\lambda_n) \notin \ell^2$. If ∂f is r-expansive, then S is a singleton $\{\hat{x}\}$ and*

$$\|x_{n+1} - \hat{x}\| \leq \|x_0 - \hat{x}\| \prod_{k=0}^{n}(1 + r^2\lambda_k^2)^{-\frac{1}{2}}.$$

In particular, (x_n) converges strongly, as $n \to \infty$, to the unique $\hat{x} \in S$.

Even Functions

Recall that a function $f : H \to \mathbf{R} \cup \{+\infty\}$ is even if $f(-x) = f(x)$ for all $x \in H$. We shall see that, when applied to an even function, the proximal point algorithm generates only strongly convergent sequences. Notice first that if f is even, then $0 \in S$, and so $S \neq \emptyset$.

Proposition 6.7. *Let (x_n) be a proximal sequence with $(\lambda_n) \notin \ell^1$. If f is even, then (x_n) converges strongly, as $n \to \infty$, to some $\hat{x} \in S$.*

Proof. Take $m > n$ and replace $u = -x_m$ in inequality (6.3) to obtain

$$\|x_{n+1} + x_m\|^2 \leq \|x_n + x_m\|^2.$$

For each fixed m, the function $n \mapsto \|x_n + x_m\|^2$ is nonincreasing. In particular,

$$4\|x_m\|^2 = \|x_m + x_m\|^2 \leq \|x_n + x_m\|^2.$$

On the other hand, the Parallelogram Identity gives

$$\|x_n + x_m\|^2 + \|x_n - x_m\|^2 = 2\|x_m\|^2 + 2\|x_n\|^2.$$

We deduce that

$$\|x_n - x_m\|^2 \le 2\|x_n\|^2 - 2\|x_m\|^2.$$

Since $\lim_{n\to\infty} \|x_n - 0\|$ exists, (x_n) must be a Cauchy sequence. $\qquad\square$

Solution Set with Nonempty Interior

If $\text{int}(S) \ne \emptyset$, there exist $\bar{u} \in S$ and $r > 0$ such that for every $h \in B(0,1)$, we have $\bar{u} + rh \in S$. Now, given $x \in \text{dom}(\partial f)$ and $x^* \in \partial f(x)$, the subdifferential inequality gives

$$f(\bar{u} + rh) \ge f(x) + \langle x^*, \bar{u} + rh - x \rangle.$$

Since $\bar{u} + rh \in S$, we deduce that

$$r\langle x^*, h \rangle \le \langle x^*, x - \bar{u} \rangle,$$

and therefore,

$$r\|x^*\| = \sup_{B(0,1)} \langle x^*, h \rangle \le \langle x^*, x - \bar{u} \rangle. \tag{6.5}$$

Proposition 6.8. *Let (x_n) be a proximal sequence with $(\lambda_n) \notin \ell^1$. If S has nonempty interior, then (x_n) converges strongly, as $n \to \infty$, to some $\hat{x} \in S$.*

Proof. If we apply Inequality (6.5) with $x = x_{n+1}$ and $x^* = \frac{x_n - x_{n+1}}{\lambda_n}$ we deduce that

$$
\begin{aligned}
2r\|x_n - x_{n+1}\| &\le 2\langle x_n - x_{n+1}, x_{n+1} - \bar{u} \rangle \\
&= \|x_n - \bar{u}\|^2 - \|x_{n+1} - \bar{u}\|^2 - \|x_{n+1} - x_n\|^2 \\
&\le \|x_n - \bar{u}\|^2 - \|x_{n+1} - \bar{u}\|^2.
\end{aligned}
$$

Whence, if $m > n$,

$$
\begin{aligned}
2r\|x_n - x_m\| &\le 2r \sum_{j=n}^{m-1} \|x_j - x_{j+1}\| \\
&\le \|x_n - \bar{u}\|^2 - \|x_m - \bar{u}\|^2.
\end{aligned}
$$

Part ii) of Proposition 6.2 implies (x_n) is a Cauchy sequence. $\qquad\square$

Linear Error Bound and Finite-Time Termination

If f is differentiable on S, the proximal point algorithm cannot terminate in a finite number of iterations unless the initial point is already in S (and the proximal sequence is stationary). Indeed, if $x_{n+1} \in S$ and f is differentiable at x_{n+1} then

$$-\frac{x_{n+1} - x_n}{\lambda_n} \in \partial f(x_{n+1}) = \{0\},$$

and so $x_{n+1} = x_n$. Proceeding inductively, one proves that $x_n = x_0 \in S$ for all n. Moreover, if the function is "too smooth" around S, one can even obtain an upper bound for the speed of convergence. To see this, let $d(u, S)$ denote the distance from u to S, and, for each $n \in \mathbf{N}$, write $\sigma_n = \sum_{k=0}^{n} \lambda_k$. We have the following:

Proposition 6.9. *Assume f is a continuously differentiable function such that ∇f is Lipschitz-continuous with constant L in a neighborhood of S and let (x_n) be a proximal sequence with $x_0 \notin S$. Then, there exists $C > 0$ such that*

$$d(x_{n+1}, S) \geq Ce^{-\sigma_n L}.$$

Proof. Recall that Px denotes the projection of x onto S. Let $\eta > 0$ be such that

$$\|\nabla f(x)\| = \|\nabla f(x) - \nabla f(Px)\| \leq L\|x - Px\| = Ld(x, S),$$

whenever $d(u, S) < \eta$. As said before, the minimizing set cannot be attained in a finite number of steps. If $d(x_N, S) < \eta$ for some N, then $d(x_{n+1}, S) < \eta$ for all $n \geq N$, and so

$$\|x_{n+1} - x_n\| = \lambda_n\|\nabla f(x_{n+1})\| \leq \lambda_n Ld(x_{n+1}, S)$$

for all $n \geq N$. But $d(x_n, S) - d(x_{n+1}, S) \leq \|x_{n+1} - x_n\|$, which implies

$$(1 + \lambda_n L)d(x_{n+1}, S) \geq d(x_n, S).$$

We conclude that

$$d(x_{n+1}, S) \geq \left[\prod_{k=N}^{n}(1 + \lambda_k L)^{-1}\right] d(x_N, S) \geq e^{-\sigma_n L} d(x_N, S),$$

and this completes the proof. $\square$

On the other hand, if the function f is somehow steep on the boundary of S, the sequence $\{x_n\}$ will reach a point in S in a finite number of iterations, which is possible to estimate in terms of the initial gap $f(x_0) - \alpha$, where $\alpha = \inf(f)$.

Proposition 6.10. *Assume $\alpha > -\infty$ and $\|v\| \geq r > 0$ for all $v \in \partial f(x)$ such that $x \notin S$. Set $N = \min\{n \in \mathbf{N} : r^2 \sigma_n \geq f(x_0) - \alpha\}$. Then $x_N \in S$.*

Proof. If $x_k \notin S$ for $k = 0, \ldots, n+1$, then

$$f(x_k) - f(x_{k+1}) \geq \left\langle -\frac{x_{k+1} - x_k}{\lambda_k}, x_k - x_{k+1}\right\rangle \geq \lambda_k \left\|\frac{x_k - x_{k+1}}{\lambda_k}\right\|^2 \geq r^2\lambda_k.$$

by the subdifferential inequality. Summing up one gets

$$r^2 \sigma_n \le f(x_0) - f(x_{n+1}) \le f(x_0) - \alpha.$$

This implies $x_N \in S$, as claimed.   □

If

$$f(x) \ge \alpha + r\|x - Px\| \tag{6.6}$$

for all $x \in H$, then

$$f(x) \ge \alpha + r\|x - Px\| \ge f(x) + \langle v, Px - x \rangle + r\|x - Px\|$$

by the subdifferential inequality. Hence $r\|x - Px\| \le \langle v, x - Px \rangle$ and so $\|v\| \ge r > 0$ for all $v \in \partial f(x)$ such that $x \notin S$. Inequality (6.6) is known as a *linear error bound*. Roughly speaking, it means that f is steep on the boundary of S.

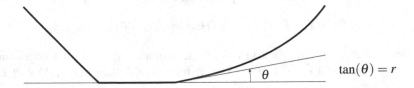

Using an equivalent form of Inequality (6.6), convergence in a finite number of iterations was proved in [93] and [55].

One remarkable property of the proximal point algorithm is that it can be applied to nonsmooth functions. Moreover, Proposition 6.10 implies that, in a sense, nonsmoothness has a positive impact in the speed of convergence.

6.2.3 Examples

In some cases, it is possible to obtain an explicit formula for the proximal iterations. In other words, for $x \in H$ and $\lambda > 0$, one can find $y \in H$ such that

$$y \in \operatorname{argmin}\{f(\xi) + \frac{1}{2\lambda}\|\xi - x\|^2 : \xi \in H\},$$

or, equivalently,

$$x - y \in \lambda \partial f(y).$$

Indicator Function

Let C be a nonempty, closed and convex subset of H. Recall that the indicator function $\delta_C : H \to \mathbf{R} \cup \{+\infty\}$ of C, given by $\delta_C(\xi) = 0$ if $\xi \in C$ and $+\infty$ otherwise, is proper, lower-semicontinuous and convex. Clearly, the set of minimizers of δ_C is precisely the set C. Take $x \in H$ and $\lambda > 0$. Then,

$$\text{argmin}\left\{ \delta_C(\xi) + \frac{1}{2\lambda}\|\xi - x\|^2 : \xi \in H \right\} = \text{argmin}\{\|\xi - x\| : \xi \in C\} = \{P_C(x)\}.$$

Observe that, in this case, the proximal iterations converge to the projection of x_0 onto $S = C$ in one iteration. The term *proximal* was introduced by J.-J. Moreau. In some sense, it expresses the fact that the proximal iteration generalizes the concept of projection when the indicator function of a set is replaced by an arbitrary proper, lower-semicontinuous and convex function (see [80], where the author presents a survey of his previous research [77, 78, 79] in the subject).

Quadratic Function

The quadratic case was studied in [71] and [72]. Consider a bounded, symmetric and positive semidefinite linear operator $A : H \to H$, and a point $b \in H$. The function $f : H \to \mathbf{R}$ defined by

$$f(x) = \frac{1}{2}\langle x, Ax \rangle - \langle b, x \rangle$$

is convex and differentiable, with gradient $\nabla f(x) = Ax - b$. Clearly, $\hat{x}$ minimizes f if, and only if, $Ax = b$. In particular, $S \neq \emptyset$ if, and only if, b belongs to the range of A. Given $x \in H$ and $\lambda > 0$, we have

$$x - y \in \lambda \partial f(y)$$

if, and only if

$$y = (I + \lambda A)^{-1}(x + \lambda b).$$

Since A is positive semidefinite, $I + \lambda A$ is indeed invertible. The iterations of the proximal point algorithm are given by

$$x_{n+1} = (I + \lambda_n A)^{-1}(x_n + \lambda_n b), \qquad n \geq 0.$$

By Theorem 6.3, if the equation $Ax = b$ has solutions, then each proximal sequence must converge weakly to one of them. Actually, the convergence is strong. To see this, take any solution $\bar{x}$ use the change of variables $u_n = x_n - \bar{x}$. A simple computation shows that

$$u_{n+1} = (I + \lambda_n A)^{-1}(u_n) = (I + \lambda_n \nabla g)^{-1}(u_n), \qquad n \geq 0,$$

where g is the *even* function given by $g(x) = \frac{1}{2}\langle x, Ax \rangle$. It follows that (u_n) converges strongly to some $\bar{u}$ and we conclude that (x_n) converges strongly to $\bar{u} + \bar{x}$.

ℓ^1 and L^1 Norms

Minimization of ℓ^1 and L^1 norms have a great number of applications in image processing and optimal control, since they are known to induce sparsity of the solutions. We shall comment these kinds of applications in the next chapter. For the moment, let us present how the proximal iterations can be computed for these functions.

In finite dimension: Let us start by considering $f_1 : \mathbf{R} \to \mathbf{R}$ defined by $f_1(\xi) = |\xi|$. Take $x \in \mathbf{R}$ and $\lambda > 0$. The unique point

$$y \in \operatorname{argmin}\left\{ |\xi| + \frac{1}{2\lambda}|\xi - x|^2 : \xi \in \mathbf{R} \right\}$$

is characterized by

$$y = \begin{cases} x - \lambda & \text{if } x > \lambda \\ 0 & \text{if } -\lambda \le x \le \lambda \\ x + \lambda & \text{if } x < -\lambda. \end{cases}$$

The preceding argument can be easily extended to the ℓ^1 norm in $\mathbf{R}^N$, namely $f_N : \mathbf{R}^N \to \mathbf{R}$ given by $f_N(\xi) = \|\xi\|_1 = |\xi_1| + \cdots + |\xi_N|$.

In $\ell^2(\mathbf{N};\mathbf{R})$: Now let $H = \ell^2(\mathbf{N};\mathbf{R})$ and define $f_\infty : H \to \mathbf{R} \cup \{+\infty\}$ by

$$f_\infty(\xi) = \|\xi\|_{\ell^1(\mathbf{N};\mathbf{R})} = \begin{cases} \sum_{i \in \mathbf{N}} |\xi_i| & \text{if } \xi \in \ell^1(\mathbf{N};\mathbf{R}) \\ +\infty & \text{otherwise.} \end{cases}$$

Take $x \in H$ and $\lambda > 0$. For each $i \in \mathbf{N}$ define

$$y_i = \begin{cases} x_i - \lambda & \text{if } x_i > \lambda \\ 0 & \text{if } -\lambda \le x_i \le \lambda \\ x_i + \lambda & \text{if } x_i < -\lambda, \end{cases}$$

Since $x \in H$ there is $I \in \mathbf{N}$ such that $|x_i| \le \lambda$ for all $i \ge I$. Hence $y_i = 0$ for all $i \ge I$ and the sequence (y_i) belongs to $\ell^1(\mathbf{N};\mathbf{R})$. Finally,

$$|y_i| + \frac{1}{2\lambda}|y_i - x_i|^2 \le |z| + \frac{1}{2\lambda}|z - x_i|^2$$

for all $z \in \mathbf{R}$ and so

$$y \in \operatorname{argmin}\left\{ \|u\|_{\ell^1(\mathbf{N};\mathbf{R})} + \frac{1}{2\lambda}\|u - x\|^2_{\ell^2(\mathbf{N};\mathbf{R})} : u \in H \right\}.$$

In $L^2(\Omega;\mathbf{R})$: Let Ω be a bounded open subset of $\mathbf{R}^N$ and consider the Hilbert space $L^2(\Omega;\mathbf{R})$, which is included in $L^1(\Omega;\mathbf{R})$. Define $f_\Omega : H \to \mathbf{R}$ by

$$f_\Omega(u) = \|u\|_{L^1(\Omega;\mathbf{R})} = \int_\Omega |u(\zeta)|\, d\zeta.$$

Take $x \in H$ (any representing function) and $\lambda > 0$. Define $y : \Omega \to \mathbf{R}$ by

$$y(\zeta) = \begin{cases} x(\zeta) - \lambda & \text{if } x(\zeta) > \lambda \\ 0 & \text{if } -\lambda \leq x(\zeta) \leq \lambda \\ x(\zeta) + \lambda & \text{if } x(\zeta) < -\lambda \end{cases}$$

for each $\zeta \in \Omega$. Observe that if one chooses a different representing function x', then the corresponding y' will match y almost everywhere. Then $y \in H$ and

$$|y(\zeta)| + \frac{1}{2\lambda}|y(\zeta) - x(\zeta)|^2 \leq |z| + \frac{1}{2\lambda}|z - x(\zeta)|^2$$

for all $z \in \mathbf{R}$. Therefore

$$y \in \operatorname{argmin}\left\{ \|u\|_{L^1(\Omega;\mathbf{R})} + \frac{1}{2\lambda}\|u - x\|_{L^2(\Omega;\mathbf{R})}^2 : u \in H \right\}.$$

6.3 Gradient-Consistent Algorithms

Gradient-consistent methods are based on the fact that if $f : H \to \mathbf{R}$ is differentiable at a point z, then $\nabla f(z)$ belongs to the normal cone of the sublevel set $\Gamma_{f(z)}(f)$. Indeed, if $y \in \Gamma_{f(z)}(f)$, we have

$$\langle \nabla f(z), y - z \rangle \leq f(y) - f(z) \leq 0$$

by the subdifferential inequality. Intuitively, a direction d forming a sufficiently small angle with $-\nabla f(z)$ must point inwards, with respect to $\Gamma_{f(z)}(f)$, and we should have $f(z + \lambda d) < f(z)$ if the step size $\lambda > 0$ is conveniently chosen.

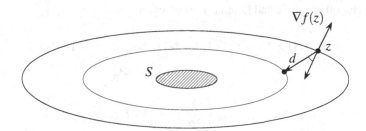

Gradient-consistent algorithms take the form

(GC)
$$\begin{cases} z_0 \in H \\ z_{n+1} = z_n + \lambda_n d_n, \end{cases}$$

where the sequence (d_n) satisfies some nondegeneracy condition with respect to the sequence of steepest descent directions, namely Hypothesis 6.12 below.

The idea (which dates back to [36]) of selecting exactly the steepest descent direction $d = -\nabla f(z)$ is reasonable, but it may not be the best option. On the other hand, the choice of the step size λ_n may be crucial in the performance of such an algorithm, as shown in the following example:

Example 6.11. Let $f : \mathbf{R} \to \mathbf{R}$ be defined by $f(z) = z^2$. Take $\lambda_n \equiv \lambda$ and set $z_0 = 1$. A straightforward computation yields $z_n = (1 - 2\lambda)^n$ for each n. Therefore, if $\lambda = 1$, the sequence (z_n) remains bounded but does not converge. Next, if $\lambda > 1$, then $\lim_{n \to \infty} z_n = +\infty$. Finally, if $\lambda < 1$, then $\lim_{n \to \infty} z_n = 0$, the unique minimizer of f. $\square$

In this section, we will explore different possibilities for λ_n and d_n, and provide qualitative and quantitative convergence results.

Hypothesis 6.12. The function f is bounded from below, its gradient ∇f is Lipschitz-continuous with constant L, there exist positive numbers α and β such that

$$\alpha \|d_n\|^2 \le \|\nabla f(z_n)\|^2 \le -\beta \langle \nabla f(z_n), d_n \rangle \tag{6.7}$$

for all $n \in \mathbf{N}$, and the sequence (λ_n) of step sizes satisfies $(\lambda_n) \notin \ell^1$ and

$$\lambda := \sup_{n \in \mathbf{N}} \lambda_n < \frac{2\alpha}{\beta L}. \tag{6.8}$$

Lemma 6.13. *Assume Hypothesis 6.12 holds. There is* $\delta > 0$ *such that*

$$f(z_{n+1}) - f(z_n) \le -\delta \lambda_n \|\nabla f(z_n)\|^2$$

for all $n \in \mathbf{N}$. *Thus, the sequence* $(f(z_n))$ *decreases to some* $v^* \in \mathbf{R}$.

Proof. Hypothesis 6.12 and Lemma 1.30 together give

$$\begin{aligned}
f(z_{n+1}) - f(z_n) &\le \langle \nabla f(z_n), z_{n+1} - z_n \rangle + \frac{L}{2} \|z_{n+1} - z_n\|^2 \\
&= \lambda_n \langle \nabla f(z_n), d_n \rangle + \frac{L\lambda_n^2}{2} \|d_n\|^2 \\
&\le \lambda_n \left[\frac{L\lambda_n}{2\alpha} - \frac{1}{\beta} \right] \|\nabla f(z_n)\|^2 \\
&\le \lambda_n \left[\frac{\beta \lambda L - 2\alpha}{2\alpha\beta} \right] \|\nabla f(z_n)\|^2.
\end{aligned}$$

It suffices to define

$$\delta = \frac{2\alpha - \beta \lambda L}{2\alpha\beta},$$

which is positive by Hypothesis 6.12. $\square$

Proposition 6.14. *Assume Hypothesis 6.12 holds. Then* $\lim\limits_{n\to\infty} \nabla f(z_n) = 0$ *and every strong limit point of the sequence* (z_n) *must be critical.*

Proof. By Lemma 6.13, we have

$$\sum_{n\in\mathbf{N}} \lambda_n \|\nabla f(z_n)\|^2 \le f(z_0) - \inf(f) < +\infty.$$

Since $(\lambda_n) \notin \ell^1$, we also have $\liminf_{n\to\infty} \|\nabla f(z_n)\| = 0$. Let $\varepsilon > 0$ and take $N \in \mathbf{N}$ large enough so that

$$\sum_{n\ge N} \lambda_n \|\nabla f(z_n)\|^2 < \frac{\varepsilon^2\sqrt{\alpha}}{4L}. \tag{6.9}$$

We shall prove that $\|\nabla f(z_m)\| < \varepsilon$ for all $m \ge N$. Indeed, if $m \ge N$ and $\|\nabla f(z_m)\| \ge \varepsilon$, define

$$v_m = \min\left\{ n > m : \|\nabla f(z_n)\| < \frac{\varepsilon}{2} \right\},$$

which is finite because $\liminf_{n\to\infty} \|\nabla f(z_n)\| = 0$. Then

$$\|\nabla f(z_m) - \nabla f(z_{v_m})\| \le \sum_{k=m}^{v_m-1} \|\nabla f(z_{k+1}) - \nabla f(z_k)\| \le L \sum_{k=m}^{v_m-1} \|z_{k+1} - z_k\|.$$

But for $k = m, \ldots, v_m - 1$ we have

$$\|z_{k+1} - z_k\| = \lambda_k \|d_k\| \le \frac{1}{\sqrt{\alpha}} \lambda_k \|\nabla f(z_k)\| \le \frac{2}{\varepsilon\sqrt{\alpha}} \lambda_k \|\nabla f(z_k)\|^2,$$

because $\|\nabla f(z_k)\| \ge \frac{\varepsilon}{2}$ for all such k. We deduce that

$$\|\nabla f(z_m)\| \le \|\nabla f(z_{v_m})\| + \frac{2L}{\varepsilon\sqrt{\alpha}} \sum_{k=m}^{v_m-1} \lambda_k \|\nabla f(z_k)\|^2 < \frac{\varepsilon}{2} + \frac{\varepsilon}{2} = \varepsilon$$

by (6.9). This contradicts the assumption that $\|\nabla f(z_m)\| \ge \varepsilon$. $\qquad\square$

If we suppose in Proposition 6.14 that the sequence (λ_n) is bounded from below by a positive constant, the proof is trivial, in view of Lemma 6.13.

So far, we have not required the objective function f to be convex.

Proposition 6.15. *If Hypothesis 6.12 holds and f is convex, then every weak limit point of the sequence (z_n) minimizes f.*

Proof. Take any $u \in H$ and let $z_{k_n} \rightharpoonup z$ as $n \to \infty$. By convexity,

$$f(u) \ge f(z_{k_n}) + \langle \nabla f(z_{k_n}), u - z_{k_n}\rangle$$

for each $n \in \mathbf{N}$. But (z_{k_n}) is bounded, $\lim\limits_{n\to\infty} \|\nabla f(z_{k_n})\| = 0$ and $\lim\limits_{n\to\infty} f(z_{k_n}) \ge f(z)$ by weak lower-semicontinuity. We conclude that $f(z) \le f(u)$. $\qquad\square$

In the strongly convex case, we immediately obtain strong convergence:

Proposition 6.16. *If Hypothesis 6.12 holds and f is strongly convex with parameter* μ, *then the sequence* (z_n) *converges strongly to the unique minimizer of f.*

Proof. Let z^* be the unique minimizer of f. The strong convexity implies

$$
\begin{aligned}
\mu \|z_{n+1} - z^*\|^2 &\leq \langle \nabla f(z_{n+1}) - \nabla f(z^*), z_{n+1} - z^* \rangle \\
&= \langle \nabla f(z_{n+1}), z_{n+1} - z^* \rangle \\
&\leq \|\nabla f(z_{n+1})\| \cdot \|z_{n+1} - z^*\|,
\end{aligned}
$$

by Proposition 3.12. It follows that $\mu \|z_{n+1} - z^*\| \leq \|\nabla f(z_{n+1})\|$. By Proposition 6.14, the right-hand side tends to 0 and so (z_n) must converge to z^*. □

Remark 6.17. Step size selection may have a great impact on the speed of convergence. Several rules have been devised and are commonly used: the *exact* and *limited minimization* rules, as well as the rules of *Armijo*, *Goldstein* and *Wolfe* (see [24]). A more sophisticated approach was developed in [81]. See also [32]. □

Let us analyze the convergence of (z_n) for specific gradient-consistent methods.

6.3.1 The Gradient Method

Let us begin by the *pure* gradient method, given by

$$
(G) \qquad \begin{cases} z_0 \in H \\ z_{n+1} = z_n - \lambda_n \nabla f(z_n) & \text{for } n \in \mathbf{N}. \end{cases}
$$

This algorithm is gradient-consistent (take $\alpha = \beta = 1$) and the condition on the step sizes is simply

$$
\sup_{n \in \mathbf{N}} \lambda_n < \frac{2}{L}.
$$

Let us recall the following basic result on real sequences:

Lemma 6.18. *Let* (a_n) *and* (ε_n) *be nonnegative sequences such that* $(\varepsilon_n) \in \ell^1$ *and* $a_{n+1} - a_n \leq \varepsilon_n$ *for each* $n \in \mathbf{N}$. *Then* $\lim_{n \to \infty} a_n$ *exists.*

Proposition 6.19. *Let* (z_n) *satisfy* (G), *where* f *is convex,* $S \neq \emptyset$, $(\lambda_n) \notin \ell^1$ *and* $\sup_{n \in \mathbf{N}} \lambda_n < \frac{2}{L}$. *Then* (z_n) *converges weakly as* $n \to \infty$ *to a point in* S.

Proof. Take $u \in S$. For each $n \in \mathbf{N}$ we have $0 \geq f(u) - f(z_n) \geq \langle \nabla f(z_n), u - z_n \rangle$. Therefore,

$$
\begin{aligned}
\|z_{n+1} - u\|^2 &= \|z_{n+1} - z_n\|^2 + \|z_n - u\|^2 + 2\langle z_{n+1} - z_n, z_n - u \rangle \\
&= \lambda_n^2 \|\nabla f(z_n)\|^2 + \|z_n - u\|^2 + 2\lambda_n \langle \nabla f(z_n), u - z_n \rangle \\
&\leq \lambda_n^2 \|\nabla f(z_n)\|^2 + \|z_n - u\|^2.
\end{aligned}
$$

Since $\sum_{n\in\mathbf{N}} \lambda_n \|\nabla f(z_n)\|^2 < \infty$, also $\sum_{n\in\mathbf{N}} \lambda_n^2 \|\nabla f(z_n)\|^2 < \infty$. We deduce the existence of $\lim_{n\to\infty} \|z_n - u\|$ by Lemma 6.18. We conclude using Proposition 6.15 and Opial's Lemma 5.2. $\qquad\qquad\square$

Using Lemma 5.3 we immediately obtain

Corollary 6.20. *Let* (z_n) *satisfy* (GC) *and assume that* $\sum_{n\in\mathbf{N}} \lambda_n \|d_n + \nabla f(z_n)\| < \infty$. *If* $(\lambda_n) \notin \ell^1$ *and* $\sup_{n\in\mathbf{N}} \lambda_n < \frac{2}{L}$, *then* (z_n) *converges weakly as* $n \to \infty$ *to a point in S.*

For strong convergence we have the following:

Proposition 6.21. *Let* (z_n) *satisfy* (G), *where* f *is convex,* $S \neq \emptyset$, $(\lambda_n) \notin \ell^1$ *and* $\sup_{n\in\mathbf{N}} \lambda_n < \frac{2}{L}$. *Then* (z_n) *converges strongly as* $n \to \infty$ *to a point in S if either* f *is strongly convex;* f *is even; or* $\mathrm{int}(\mathrm{argmin}(f)) \neq \emptyset$.

6.3.2 Newton's Method

Let $f : H \to \mathbf{R} \cup \{+\infty\}$ and let $z^* \in S$. Assume that f is twice Gâteaux-differentiable in a neighborhood of z^*, that its Hessian $\nabla^2 f$ is continuous in z^*, and that the operator $\nabla^2 f(z^*)$ is invertible. By Proposition 1.8, there exist a convex neighborhood U of z^* and a constant $C > 0$ such that f is strictly convex on U, $\nabla^2 f(z)$ is invertible for all $z \in U$, and

$$\sup_{z\in U} \|[\nabla^2 f(z)]^{-1}\|_{\mathscr{L}(H;H)} \leq C.$$

Newton's Method is defined by

$$(NM) \qquad \begin{cases} z_0 \in U \\ z_{n+1} = z_n - [\nabla^2 f(z_n)]^{-1} \nabla f(z_n) \quad \text{for } n \in \mathbf{N}. \end{cases}$$

Example 6.22. If f is quadratic, namely

$$f(x) = \langle Ax, x \rangle + \langle b, x \rangle + c,$$

then $\nabla^2 f(x) = A + A^*$ for all $x \in U = H$. If the operator $A + A^*$ is invertible, then Newton's Method can be applied and converges in *one* iteration. Indeed, a straightforward computation shows that $\nabla f(z_1) = 0$. We should point out that, in this setting, computing this one iteration is equivalent to solving the optimality condition $\nabla f(x) = 0$. $\qquad\qquad\square$

In general, if the initial point z_0 is sufficiently close to z^*, Newton's method converges very fast. More precisely, we have the following:

Proposition 6.23. *Let* (z_n) *be defined by Newton's Method* (NM) *with* $z_0 \in U$. *We have the following:*

i) *For each $\varepsilon \in (0,1)$, there is $R > 0$ such that, if $\|z_N - z^*\| \leq R$ for some $N \in \mathbf{N}$, then*

$$\|z_{N+m} - z^*\| \leq R\varepsilon^m$$

for all $m \in \mathbf{N}$.

ii) *Assume $\nabla^2 f$ is Lipschitz-continuous on U with constant M and let $\varepsilon \in (0,1)$. If $\|z_N - z^*\| \leq \frac{2\varepsilon}{MC}$ for some $N \in \mathbf{N}$, then*

$$\|z_{N+m} - z^*\| \leq \frac{2}{MC}\varepsilon^{2^m}$$

for all $m \in \mathbf{N}$.

Proof. Fix $n \in N$ with $z_n \in U$, and define $g : [0,1] \to H$ by $g(t) = \nabla f(z^* + t(z_n - z^*))$. We have

$$\nabla f(z_n) = g(1) - g(0) = \int_0^1 \dot{g}(t)\, dt = \int_0^1 \nabla^2 f(z^* + t(z_n - z^*))(z_n - z^*)\, dt.$$

It follows that

$$\|z_{n+1} - z^*\| = \left\| z_n - z^* - [\nabla^2 f(z_n)]^{-1} \nabla f(z_n) \right\|$$

$$= \left\| [\nabla^2 f(z_n)]^{-1} \left[\nabla^2 f(z_n)(z_n - z^*) - \nabla f(z_n) \right] \right\|$$

$$\leq C \left\| \int_0^1 \left[\nabla^2 f(z_n) - \nabla^2 f(z^* + t(z_n - z^*)) \right] (z_n - z^*)\, dt \right\|$$

$$\leq C\|z_n - z^*\| \int_0^1 \left\| \nabla^2 f(z_n) - \nabla^2 f(z^* + t(z_n - z^*)) \right\|_{\mathscr{L}(H;H)}\, dt \quad (6.10)$$

Take $\varepsilon \in (0,1)$ and pick $R > 0$ such that $\|\nabla^2 f(z) - \nabla^2 f(z^*)\| \leq \varepsilon/2C$ whenever $\|z - z^*\| \leq R$. We deduce that if $\|z_N - z^*\| < R$, then $\|z_{n+1} - z^*\| \leq \varepsilon\|z_n - z^*\|$ for all $n \geq N$, and this implies i).

For part ii), using (6.10) we obtain $\|z_{n+1} - z^*\| \leq \frac{MC}{2}\|z_n - z^*\|^2$. Given $\varepsilon \in (0,1)$, if $\|z_N - z^*\| \leq \frac{2\varepsilon}{MC}$, then $\|z_{N+m} - z^*\| \leq \frac{2}{MC}\varepsilon^{2^m}$ for all $m \in \mathbf{N}$. $\qquad\square$

Despite its remarkable local speed of convergence, Newton's method has some noticeable drawbacks:

First, the fast convergence can only be granted once the iterates are sufficiently close to a solution. One possible way to overcome this problem is to perform a few iterations of another method in order to provide a *warm* starting point.

Next, at points where the Hessian is close to degeneracy, the method may have a very erratic behavior.

Example 6.24. Consider the function $f : \mathbf{R} \to \mathbf{R}$ be defined by $f(z) = \sqrt{z^2 + 1}$. A simple computation shows that Newton's method gives $z_{n+1} = -z_n^3$ for each n. The problem here is that, since f is almost flat for large values of z_n, the method predicts that the minimizers should be very far. □

Some alternatives are the following:

- Adding a small constant λ_n (a step size), and computing

$$z_{n+1} = z_n - \lambda_n [\nabla^2 f(z_n)]^{-1} \nabla f(z_n)$$

 can help to avoid very large displacements. The constant λ_n has to be carefully selected: If it is too large, it will not help; while, if it is too small, convergence may be very slow.
- Another option is to force the Hessian away from degeneracy by adding a uniformly elliptic term, and compute

$$z_{n+1} = z_n - [\nabla^2 f(z_n) + \varepsilon I]^{-1} \nabla f(z_n),$$

where $\varepsilon > 0$ and I denotes the identity operator.

Finally, each iteration has a high computational complexity due to the inversion of the Hessian operator. A popular solution for this is to update the Hessian every once in a while. More precisely, compute $[\nabla^2 f(z_N)]^{-1}$ and compute

$$z_{n+1} = z_n - \lambda_n [\nabla^2 f(z_N)]^{-1} \nabla f(z_n)$$

for the next p_N iterations. After that, one computes $[\nabla^2 f(z_{N+p_N})]^{-1}$ and continues iterating in a similar fashion.

6.4 Some Comments on Extensions and Variants

Many constrained problems can be reinterpreted, reduced or simplified, so they can be more efficiently solved. Moreover, in some cases, this preprocessing is even necessary, in order to make problems computationally tractable.

6.4.1 Additive Splitting: Proximal and Gradient Methods

The proximal point algorithm is suitable for solving minimization problems that involve nonsmooth objective functions, but can be hard to implement. If the objective function is sufficiently regular, it may be more effective and efficient to use gradient-consistent methods, which are easier to implement; or Newton's method, which converges remarkably fast. Let $f = f_1 + f_2$. If problem involves smooth and nonsmooth features (if f_1 is smooth but f_2 is not), it is possible to combine these approaches and construct better adapted methods. If neither f_1 nor f_2 is smooth, it may still be useful to treat them separately since it may be much harder to implement the proximal iterations for the sum $f_1 + f_2$, than for f_1 and f_2 independently. A thorough review of these methods can be found in [21].

Forward–Backward, Projected Gradient and Tseng's Algorithm

For instance, suppose that $f_1 : X \to \mathbf{R}$ is smooth, while $f_2 : X \to \mathbf{R} \cup \{+\infty\}$ is only proper, convex and lower-semicontinuous. We have seen in previous sections that the proximal point algorithm is suitable for the minimization of f_2, but harder to implement; while the gradient method is simpler and effective for the minimization of f_1. One can combine the two methods as:

$$x_{n+1} = \gamma(I + \lambda_n \partial f_2)^{-1}(x_n - \lambda_n \nabla f_1(x_n)) + (1-\gamma)x_n, \qquad \gamma \in (0,1]^1$$

This scheme is known as the *forward–backward* algorithm. It is a generalization of the *projected gradient* method introduced by [59] and [73], which corresponds to the case $f_2 = \delta_C$, where C is a nonempty, closed and convex subset of H, and reads

$$x_{n+1} = P_C(x_n - \lambda_n \nabla f_1(x_n)).$$

Example 6.25. For the structured problem with affine constraints

$$\min\{\phi(x) + \psi(y) : Ax + By = c\},$$

the projected gradient method reads

$$(x_{n+1}, y_{n+1}) = P_V\left(x_n - \lambda_n \nabla \phi(x_n), y_n - \lambda_n \nabla \psi(y_n)\right),$$

where $V = \{(x,y) : Ax + By = c\}$. □

Example 6.26. For the problem

$$\min\{\gamma\|v\|_1 + \|Av - b\|^2 : v \in \mathbf{R}^n\},$$

the forward–backward algorithm gives the *iterative shrinkage/thresholding algorithm* (ISTA):

$$x_{n+1} = (I + \lambda \gamma \|\cdot\|_1)^{-1}(x_n - 2\lambda A^*(Ax_n - b)),$$

where

$$\left((I + \lambda \gamma \|\cdot\|_1)^{-1}(v)\right)_i = (|v_i| - \lambda \gamma)^+ \operatorname{sgn}(v_i), \ i = 1,\ldots,n$$

(see [22, 23, 42]). See also [104] □

A variant of the forward–backward algorithm was proposed by Tseng [101] for $\min\{f_1(x) + f_2(x) : x \in C\}$. It includes a second gradient subiteration and closes

---

[1] Using $\gamma \in (0,1)$ is sometimes useful for stabilization purposes.

with a projection onto C:

$$\begin{cases} y_n = x_n - \lambda_n \nabla f_1(x_n) \\ z_n = (I + \lambda_n \partial f_2)^{-1}(y_n) \\ w_n = z_n - \lambda_n \nabla f_1(z_n) \\ x_{n+1} = P_C(x_n + (z_n - y_n)). \end{cases}$$

Double-Backward, Alternating Projections and Douglas-Rachford Algorithm

If neither f_1 nor f_2 is smooth, an alternative is to use a *double-backward* approach:

$$x_{n+1} = \gamma(I + \lambda_n \partial f_2)^{-1}(I + \lambda_n \partial f_1)^{-1}(x_n) + (1 - \gamma)x_n, \qquad \gamma \in (0, 1].$$

Example 6.27. For example, in order to find the intersection of two closed convex sets, C_1 and C_2, this gives (with $\gamma = 1$) the *method of alternating projections*:

$$x_{n+1} = P_{C_2} P_{C_1}(x_n),$$

studied in [40, 61, 102], among others. As one should expect, sequences produced by this method converge weakly to points in $C_1 \cap C_2$. Convergence is strong if C_1 and C_2 are closed, affine subspaces.   □

A modification of the double-backward method, including an over-relaxation between the backward subiterations, gives the *Douglas-Rachford* [49] algorithm:

$$\begin{cases} y_n = (I + \lambda_n \partial f_1)^{-1}(x_n) \\ z_n = (I + \lambda_n \partial f_2)^{-1}(2y_n - x_n) \\ x_{n+1} = x_n + \gamma(z_n - y_n), \qquad \gamma \in (0, 2). \end{cases}$$

6.4.2 Duality and Penalization

Multiplier Methods

Let $f, g : H \to \mathbf{R} \cup \{+\infty\}$ be proper, convex and lower-semicontinuous, and recall from Sect. 3.7 that solutions for $\min\{f(x) : g(x) \leq 0\}$ are saddle points of the Lagrangian $L(x, \mu) = f(x) + \mu g(x)$. This suggests that an iterative procedure based on minimization with respect to the primal variable x, and maximization with respect to the dual variable μ, may be an effective strategy for solving this problem. Several methods are based on the *augmented Lagrangian*

$$L_r(x, \mu) = f(x) + \mu g(x) + rg(x)^2,$$

where $r > 0$. This idea was introduced in [63] and [89], and combined with proximal iterations in [92]. Similar approaches can be used for the affinely constrained problem $\min\{f(x) : Ax = b\}$, whose Lagrangian is $L(x, y^*) = f(x) + \langle y^*, Ax - b \rangle$. Moreover, for a structured problem $\min\{f_1(x) + f_2(y) : Ax + By = c\}$, one can use alternating methods of multipliers. This approach has been followed by [15, 103] and [38] (see also [52, 62] and [37]). In the latter, a multiplier prediction step is added and implementability is enhanced. See also [27, 97], and the references therein, for further commentaries on augmented Lagrangian methods.

Penalization

Recall, from Remark 5.4, that the penalization technique is based on the idea of progressively replacing $f + \delta_C$ by a family (f_ν), in such a way that the minimizers of f_ν approximate the minimizers of $f + \delta_C$, as $\nu \to \nu_0$. Iterative procedures can be combined with penalization schemes in a *diagonal* fashion. Starting from a point $x_0 \in H$, build a sequence (x_n) as follows: At the n-th iteration, you have a point x_n. Choose a value ν_n for the penalization parameter, and apply one iteration of the algorithm of your choice (for instance, the proximal point algorithm or the gradient method), to obtain x_{n+1}. Update the value of your parameter to ν_{n+1}, and iterate.

Some interior penalties: Let $f, g : H \to \mathbf{R}$ be convex (thus continuous), and set $C = \{x \in H : g(x) \le 0\}$. Consider the problem $\min\{f(x) : x \in C\}$. Let $\theta : \mathbf{R} \to \mathbf{R} \cup \{+\infty\}$ be a convex, increasing function such that $\lim_{u \to 0} \theta(u) = +\infty$ and $\lim_{u \to -\infty} \frac{\theta(u)}{u} = 0$. For $\nu > 0$, define $f_\nu : H \to \mathbf{R} \cup \{+\infty\}$ by $f_\nu(x) = f(x) + \nu\theta(g(x)/\nu)$. Then $\mathrm{dom}(f_\nu) = \{x \in H : g(x) < 0\} = \mathrm{int}(C)$ for all $\nu > 0$, and $\lim_{\nu \to 0} f_\nu(x) = f(x)$ for all $x \in \mathrm{int}(C)$.

Example 6.28. The following functions satisfy the hypotheses described above:

1. *Logarithmic barrier*: $\theta(u) = -\ln(-u)$ for $u < 0$ and $\theta(u) = +\infty$ if $u \ge 0$.
2. *Inverse barrier*: $\theta(u) = -1/u$ for $u < 0$ and $\theta(u) = +\infty$ if $u \ge 0$. □

In [43], the authors consider a number of penalization schemes coupled with the proximal point algorithm (see also [2, 16] and the references in [43]). Other related approaches include [17, 65] and [51].

Exterior penalties: Let $f : H \to \mathbf{R}$ be convex and let Ψ be an exterior penalization function for C. Depending on the regularity of f and Ψ, possibilities for solving $\min\{f(x) : x \in C\}$ include proximal and double backward methods ([12]):

$$x_{n+1} = (I + \lambda_n \partial f + \lambda_n \beta_n \partial \Psi)^{-1}(x_n), \text{ or } x_{n+1} = (I + \lambda_n \partial f)^{-1}(I + \lambda_n \beta_n \partial \Psi)^{-1}(x_n),$$

a forward–backward algorithm ([13] and [83]):

$$x_{n+1} = (I + \lambda_n \partial f)^{-1}\left(x_n - \lambda_n \beta_n \nabla \Psi(x_n)\right),$$

or a gradient method ([86]):

$$x_{n+1} = x_n - \lambda_n \nabla f(x_n) - \lambda_n \beta_n \nabla \Psi(x_n).$$

For instance, if $C = \{x \in H : Ax = b\}$, we can write $\Psi(x) - \frac{1}{2}\|Ax - b\|_K^2$, so that $\nabla\Psi(x) = A^*(Ax_n - b)$ and $\lim_{\beta \to +\infty} \beta\Psi(x) = \delta_C(x)$.

Alternating variants have been studied in [8, 11, 14], among others. See also [57], where the penalization parameter is interpreted as a Lagrange multiplier. Related methods using an exponential penalization function have been studied, for instance, in [3, 4, 44] and [75].

References

1. Adams R. Sobolev spaces. Academic; 1975.
2. Alart P, Lemaire B. Penalization in nonclassical convex programming via variational convergence. Math Program. 1991;51:307–31.
3. Álvarez F. Absolute minimizer in convex programming by exponential penalty. J Convex Anal. 2000;7:197–202.
4. Alvarez F, Cominetti R. Primal and dual convergence of a proximal point exponential penalty method for linear programming. Math Program. 2002;93:87–96.
5. Álvarez F, Peypouquet J. Asymptotic equivalence and Kobayashi-type estimates for nonautonomous monotone operators in Banach spaces. Discret Contin Dynam Syst. 2009;25:1109–28.
6. Álvarez F, Peypouquet J. Asymptotic almost-equivalence of Lipschitz evolution systems in Banach spaces. Nonlinear Anal. 2010;73:3018–33.
7. Álvarez F, Peypouquet J. A unified approach to the asymptotic almost-equivalence of evolution systems without Lipschitz conditions. Nonlinear Anal. 2011;74:3440–4.
8. Attouch H, Bolte J, Redont P, Soubeyran A. Alternating proximal algorithms for weakly coupled convex minimization problems, Applications to dynamical games and PDE's. J Convex Anal. 2008;15:485–506.
9. Attouch H, Brézis H. Duality for the sum of convex functions in general Banach spaces. In: Barroso J, editor. Aspects of mathematics and its applications. Amsterdam: North-Holland Publishing Co; 1986. pp. 125–33.
10. Attouch H, Butazzo G, Michaille G. Variational analysis in Sobolev and BV spaces: applications to PDEs and optimization. Philadelphia: SIAM; 2005.
11. Attouch H, Cabot A, Frankel P, Peypouquet J. Alternating proximal algorithms for constrained variational inequalities. Application to domain decomposition for PDE's. Nonlinear Anal. 2011;74:7455–73.
12. Attouch H, Czarnecki MO, Peypouquet J. Prox-penalization and splitting methods for constrained variational problems. SIAM J Optim. 2011;21:149–73.
13. Attouch H, Czarnecki MO, Peypouquet J. Coupling forward-backward with penalty schemes and parallel splitting for constrained variational inequalities. SIAM J Optim. 2011;21:1251–74.
14. Attouch H, Redont P, Soubeyran A. A new class of alternating proximal minimization algorithms with costs-to-move. SIAM J Optim. 2007;18:1061–81.
15. Attouch H, Soueycatt M. Augmented Lagrangian and proximal alternating direction methods of multipliers in Hilbert spaces. Applications to games, PDE's and control. Pacific J Opt. 2009;5:17–37.
16. Auslender A, Crouzeix JP, Fedit P. Penalty-proximal methods in convex programming. J Optim Theory Appl. 1987;55:1–21.

17. Auslender A, Teboulle M. Interior gradient and proximal methods for convex and conic optimization. SIAM J Optim. 2006;16:697–725.
18. Bachmann G, Narici L, Beckenstein E. Fourier and Wavelet Analysis. New York: Springer; 2002.
19. Baillon JB. Comportement asymptotique des contractions et semi-groupes de contraction. Thèse. Université Paris 6; 1978.
20. Baillon JB. Un exemple concernant le comportement asymptotique de la solution du problème $du/dt + \partial \varphi(u) \ni 0$. J Funct Anal. 1978;28:369–76.
21. Bauschke H, Combettes PL. Convex analysis and monotone operator theory in Hilbert spaces. New York: Springer; 2011.
22. Beck A, Teboulle M. A fast iterative shrinkage-thresholding algorithm for linear inverse problems. SIAM J Imaging Sci. 2009;2:183–202.
23. Beck A, Teboulle M. Gradient-based algorithms with applications in signal recovery problems. Convex optimization in signal processing and communications. Cambridge: Cambridge University Press; 2010. pp. 33–88.
24. Bertsekas D. Nonlinear programming. Belmont: Athena Scientific; 1999.
25. Borwein J. A note on ε-subgradients and maximal monotonicity. Pac J Math. 1982;103:307–14.
26. Borwein J, Lewis A. Convex analysis and nonlinear optimization. New York: Springer; 2010.
27. Boyd S, Parikh N, Chu E, Peleato B, Eckstein J. Distributed optimization and statistical learning via the alternating direction method of multipliers. Found Trends Mach Learn. 2011;3:1–122.
28. Brenner SC, Scott LR. The mathematical theory of finite element methods. 3rd ed. New York: Springer; 2008.
29. Brézis H. Opérateurs maximaux monotones et semi-groupes de contractions dans les espaces de Hilbert. Amsterdam: North Holland Publishing Co; 1973.
30. Brézis H. Functional analysis, Sobolev spaces and partial differential equations. New York: Springer; 2011.
31. Brézis H, Lions PL. Produits infinis de résolvantes. Israel J Math. 1978;29:329–45.
32. Burachik R, Grana Drummond LM, Iseum AN, Svaiter BF. Full convergence of the steepest descent method with inexact line searches. Optimization. 1995;32:137–46.
33. Cartan H. Differential calculus. London: Kershaw Publishing Co; 1971.
34. Candès EJ, Romberg JK, Tao T. Robust uncertainty principles: exact signal reconstruction from highly incomplete frequency information. IEEE Trans Inform Theory. 2006;52:489–509.
35. Candès EJ, Romberg JK, Tao T. Stable signal recovery from incomplete and inaccurate measurements. Comm Pure Appl Math. 2006;59:1207–23.
36. Cauchy AL. Méthode générale pour la résolution des systèmes d'équations simultanées. C R Acad Sci Paris. 1847;25:536–8.
37. Chambolle A, Pock T. A first-order primal-dual algorithm for convex problems with applications to imaging. J Math Imaging Vis. 2010;40:1–26.
38. Chen G, Teboulle M. A proximal-based decomposition method for convex minimization problems. Math Program. 1994;64:81–101.
39. Chen SS, Donoho DL, Saunders MA. Atomic decomposition by basis pursuit. SIAM J Sci Comput. 1998;20:33–61.
40. Cheney W, Goldstein AA. Proximity maps for convex sets. Proc Amer Math Soc. 1959;10:448–50.
41. Ciarlet PG. The finite element method for elliptic problems. Amsterdam: North Holland Publishing Co; 1978.
42. Combettes PL, Wajs VR. Signal recovery by proximal forward-backward splitting. Multiscale Model Simul. 2005;4:1168–200.
43. Cominetti R, Courdurier M. Coupling general penalty schemes for convex programming with the steepest descent method and the proximal point algorithm. SIAM J Opt. 2002;13:745–65.
44. Cominetti R, San Martín J. Asymptotic analysis of the exponential penalty trajectory in linear programming. Math Program. 1994;67:169–87.

45. Crandall MG, Liggett TM. Generation of semigroups of nonlinear transformations on general Banach spaces. Am J Math. 1971;93:265–98.
46. Daubechies I. Ten lectures on wavelets. Philadelphia: SIAM; 1992.
47. Donoho DL. Compressed sensing. IEEE Trans Inform Theory. 2006;52:1289–306.
48. Donoho DL, Tanner J. Sparse nonnegative solutions of underdetermined linear equations by linear programming. Proc Natl Acad Sci U S A. 2005;102:9446–51.
49. Douglas J Jr, Rachford HH Jr. On the numerical solution of heat conduction problems in two and three space variables. Trans Am Math Soc. 1956;82:421–39.
50. Dunford N, Schwartz J. Linear operators. Part I. General theory. New York: Wiley; 1988.
51. Eckstein J. Nonlinear proximal point algorithms using Bregman functions, with applications to convex programming. Math Oper Res. 1993;18:202–26.
52. Eckstein J. Some saddle-function splitting methods for convex programming. Opt Methods Softw. 1994;4:75–83.
53. Ekeland I, Temam R. Convex analysis and variational problems. Philadelphia: SIAM; 1999.
54. Evans L. Partial differential equations. 2nd edn. Providence: Graduate Studies in Mathematics, AMS; 2010.
55. Ferris M. Finite termination of the proximal point algorithm. Math Program. 1991;50:359–66.
56. Folland G. Fourier analysis and its applications. Pacific Grove: Wadsworth & Brooks/Cole; 1992.
57. Frankel P, Peypouquet J. Lagrangian-penalization algorithm for constrained optimization and variational inequalities. Set-Valued Var Anal. 2012;20:169–85.
58. Friedlander MP, Tseng P. Exact regularization of convex programs. SIAM J Optim. 2007;18:1326–50.
59. Goldstein AA. Convex programming in Hilbert space. Bull Am Math Soc. 1964;70:709–10.
60. Güler O. On the convergence of the proximal point algorithm for convex minimization. SIAM J Control Optim. 1991;29:403–19.
61. Halperin I. The product of projection operators. Acta Sci Math (Szeged). 1962;23:96–9.
62. He B, Liao L, Han D, Yang H. A new inexact alternating directions method for monotone variational inequalities. Math Program. 2002;92:103–18.
63. Hestenes MR. Multiplier and gradient methods. J Opt Theory Appl. 1969;4:303–20.
64. Hiriart-Urruty JB, Lemaréchal C. Fundamentals of convex analysis. New York: Springer; 2001.
65. Iusem A, Svaiter BF, Teboulle M. Entropy-like proximal methods in convex programming. Math Oper Res. 1994;19:790–814.
66. Izmailov A, Solodov M. Optimization, Volume 1: optimality conditions, elements of convex analysis and duality. Rio de Janeiro: IMPA; 2005.
67. Izmailov A, Solodov M. Optimization, Volume 2: computational methods. Rio de Janeiro: IMPA; 2007.
68. James R. Reflexivity and the supremum of linear functionals. Israel J Math. 1972;13:289–300.
69. Kinderlehrer D, Stampacchia G. An introduction to variational inequalities and their applications. New York: Academic; 1980.
70. Kobayashi Y. Difference approximation of Cauchy problems for quasi-dissipative operators and generation of nonlinear semigroups. J Math Soc Japan. 1975;27:640–65.
71. Krasnosel'skii MA. Solution of equations involving adjoint operators by successive approximations. Uspekhi Mat Nauk. 1960;15:161–5.
72. Kryanev AV. The solution of incorrectly posed problems by methods of successive approximations. Dokl Akad Nauk SSSR. 1973;210:20–2. Soviet Math Dokl. 1973;14:673–6.
73. Levitin ES, Polyak BT. Constrained minimization methods. USSR Comp Math Math Phys. 1966;6:1–50.
74. Lions JL. Quelques méthodes de résolution des problèmes aux limites non linéaires. Paris: Dunod; 2002.
75. Mangasarian OL, Wild EW. Exactness conditions for a convex differentiable exterior penalty for linear programming. Optimization. 2011;60:3–14.

76. Martinet R. Régularisation d'inéquations variationnelles par approximations successives. Rev Française Informat Recherche Opérationnelle. 1970;4:154–8.

77. Moreau JJ. Décomposition orthogonale dun espace hilbertien selon deux cônes mutuellement polaires. C R Acad Sci Paris. 1962;255:238–40.

78. Moreau JJ. Fonctions convexes duales et points proximaux dans un espace hilbertien. C R Acad Sci Paris. 1962;255:2897–9.

79. Moreau JJ. Proprietés des applications "prox". C R Acad Sci Paris. 1963;256:1069–71.

80. Moreau JJ. Proximité et dualité dans un espace hilbertien. Bull Soc Math France. 1965;93:273–99.

81. Nesterov YE. A method for solving the convex programming problem with convergence rate $O(1/k^2)$. Dokl Akad Nauk SSSR. 1983;269:543–7.

82. Nocedal J, Wright S. Numerical optimization. 2nd ed. New York: Springer; 2006.

83. Noun N, Peypouquet J. Forward-backward-penalty scheme for constrained convex minimization without inf-compactness. J Optim Theory Appl. 2013;158:787–95.

84. Opial Z. Weak Convergence of the sequence of successive approximations for nonexpansive mappings. Bull Amer Math Soc. 1967;73:591-7.

85. Peypouquet J. Asymptotic convergence to the optimal value of diagonal proximal iterations in convex minimization. J Convex Anal. 2009;16:277–86.

86. Peypouquet J. Coupling the gradient method with a general exterior penalization scheme for convex minimization. J Optim Theory Appl. 2012;153:123–38.

87. Peypouquet J, Sorin S. Evolution equations for maximal monotone operators: asymptotic analysis in continuous and discrete time. J Convex Anal. 2010;17:1113–63.

88. Peypouquet J. Optimización y sistemas dinámicos. Caracas: Ediciones IVIC; 2013.

89. Powell MJD. A method for nonlinear constraints in minimization problems. In: Fletcher R, editor. Optimization. New York: Academic; 1969. pp. 283–98.

90. Raviart PA, Thomas JM. Introduction à l'analyse numérique des équations aux dérivées partielles. Paris: Masson; 1983.

91. Rockafellar RT. Convex analysis. Princeton: Princeton University Press; 1970.

92. Rockafellar RT. Augmented lagrangians and applications of the proximal point algorithm in convex programming. Math Oper Res. 1976;1:97–116.

93. Rockafellar RT. Monotone operators and the proximal point algorithm. SIAM J Control Optim. 1976;14:877–98.

94. Rudin W. Functional analysis. Singapore: McGraw-Hill Inc; 1991.

95. Sagastizábal C, Solodov M. An infeasible bundle method for nonsmooth convex constrained optimization without a penalty function or a filter. SIAM J Optim. 2005;16:146–69.

96. Schaefer HH, Wolff MP. Topological vector spaces. New York: Springer; 1999.

97. Shefi R, Teboulle M. Rate of convergence analysis of decomposition methods based on the proximal method of multipliers for convex minimization. SIAM J Optim. 2014;24:269–97.

98. Solodov MV, Svaiter BF. Forcing strong convergence of proximal point iterations in a Hilbert space. Math Program. 2000;87:189–202.

99. Tibshirani R. Regression shrinkage and selection via the Lasso. J Royal Stat Soc Ser B. 1996;58:267–88.

100. Tolstov GP. Fourier series. Mineola: Dover; 1976.

101. Tseng P. A modified forward-backward splitting method for maximal monotone mappings. SIAM J Control Optim. 2000;38:431–46.

102. von Neumann J. On rings of operators. Reduction theory. Ann Math. 1949;50:401–85.

103. Xu MH. Proximal alternating directions method for structured variational inequalities. J Optim Theory Appl. 2007;134:107–17.

104. Yang J, Zhang Y. Alternating direction algorithms for ℓ^1-problems in compressive sensing. SIAM J Sci Comp. 2011;33:250–78.

105. Zalinescu C. Convex analysis in general vector spaces. Singapore: World Scientific; 2002.

Index

J. Peypouquet, *Convex Optimization in Normed Spaces*,
SpringerBriefs in Optimization, DOI 10.1007/978-3-319-13710-0